Russia's Factory
CHILDREN

**WITHDRAWN
UTSA LIBRARIES**

Pitt Series in Russian and East European Studies

Jonathan Harris, Editor

Russia's Factory
CHILDREN

State, Society, and Law, 1800–1917

BORIS B. GORSHKOV

UNIVERSITY OF PITTSBURGH PRESS

Published by the University of Pittsburgh Press, Pittsburgh, Pa., 15260
Copyright © 2009, University of Pittsburgh Press
All rights reserved
Manufactured in the United States of America
Printed on acid-free paper

10   9   8   7   6   5   4   3   2   1

Library of Congress Cataloging-in-Publication Data
Gorshkov, Boris B.
Russia's factory children : state, society, and law, 1800–1917 / Boris B. Gorshkov.
    p. cm. — (Pitt series in Russian and East European studies)
Includes bibliographical references and index.
ISBN-13: 978-0-8229-4383-9 (cloth : alk. paper)
ISBN-10: 0-8229-4383-2 (cloth : alk. paper)
ISBN-13: 978-0-8229-6048-5 (pbk. : alk. paper)
ISBN-10: 0-8229-6048-6 (pbk. : alk. paper)
1. Child labor—Russia—History—19th century.   2. Manufacturing industries—Employees—Supply and demand—Russia—History—19th century.   3. Industrialization—Russia—History—19th century. 4. Children—Russia—Social conditions—19th century.   I. Title.
HD6250.R92G67 2009
331.3´1094709034—dc22   2009024344

To Stepashka and the entire company

Contents

	Acknowledgments	ix
	Introduction: The Problem and the Context	1
1	Origins of Child Industrial Labor	12
2	Children in Industry: Demographic and Social Context	46
3	Public Debates and Legislative Efforts	93
4	Factory Children: Politics, Education, and the State	128
	Conclusion: Experience and Outcome	174
	Appendix: Documents	181
	Notes	185
	Bibliography	201
	Index	211

Illustrations follow page 86

Acknowledgments

THIS STUDY WAS FIRST undertaken at Auburn University. I am grateful to all members of the Auburn University History Department, who helped me enter the profession and gave me intellectual support and stimulus. I carried out preliminary research on the topic of child labor in Moscow and St. Petersburg with the help of generous travel grants from Auburn University. I am thankful to the staffs of the State Archive of the Russian Federation (GARF), the Central Historical Archive of Moscow (TsIAM), the Russian State Archive of Ancient Acts (RGADA), the Russian State Historical Archive (RGIA), the Russian Public Library, and the State Historical Library. I would like to thank professors Jane Burbank of New York University and Dan Szechi of the University of Manchester for their intellectual encouragement and support. Most especially I wish to express my gratitude to Professor Michael Melancon for more than a decade of inspiration and for repeatedly helping to light the way. Various parts of this book have been presented at conferences, and I am thankful to all panel participants and discussants, who gave me many helpful comments, suggestions, and encouragement. I am also grateful to Peter Kracht, the editorial director of the University of Pittsburgh Press, for his constant support and encouragement and to the manuscript's reviewers for a multitude of helpful suggestions. My final thanks go to my mother, who patiently accepted my presence in her home in Moscow during innumerable research trips and who always gave me love and support.

Russia's Factory
CHILDREN

Introduction

The Problem and the Context

In childhood's golden times,
Everyone lives happily—
Effortless and lighthearted
With fun and joy.
Only we don't get to run and play
in the golden fields:
All day the factory's wheels
We turn, and turn, and turn . . .

           N. A. Nekrasov, "Children's Cry"

THE PASSAGE FROM NEKRASOV'S POEM captures the harsh realities of child labor in nineteenth-century Russian factories.[1] Child industrial labor outraged many great writers of the era, including Anton Chekhov, Maxim Gorky, and Fyodor Dostoevsky.[2] A late nineteenth-century observer wrote that in order "to see the conditions of children in the mines, one needs to enter the machine plant, or the lamp workshop, where the atmosphere is suffused with the smell of gasoline used for lamps, which causes headache and nausea. Inside [the mine] one can see an entire chain of small boys, moving around the gasoline lamps, wiping and fueling them."[3] Child labor also drew the attention of great contemporary artists and painters, such as V. E. Makovskii and Il'ya Repin.

Regardless of the hardship involved, children in Russia, as elsewhere, labored in industries of all types. The extent of children's employment suggests the enormous role children played in the development of the Russian industrial economy. Children made up a surprisingly large segment of the industrial labor force. Most working children were rural residents who came to industrial areas with their parents or relatives or were recruited in the countryside by employers. A few were urban children of poor families or inmates of foundling homes. Throughout the country, industries usually employed children in unskilled and auxiliary tasks. In sugar plants, they worked inside boilers, scaling and cleaning them. In mines, children fueled kerosene lamps and carried mining equipment. On occasion, children even performed tasks normally done by adult workers. In the textile industry, for instance, children commonly assisted adult workers by carrying bobbins and cleaning equipment and floors but also sometimes worked as spinners and weavers.

In the mid-nineteenth century, the average number of children aged sixteen and under employed in industry accounted for about 15 percent of all Russia's industrial workers, varying in individual businesses, however, from 0 to 40 percent. With the rapid development of the economy during the following decades, industry's reliance on child labor intensified. Industries remunerated the labor of children they employed at one-third the lowest rate of the typical adult male worker. The children's workday lasted for twelve and even more hours. Deprived of their childhood, factory children learned early on all the responsibilities and grievances of adult life. They shared all burdens with their parents and became an important element of the family economy. Because of the hardship involved, by the late nineteenth century, child labor had become a matter of serious concern for many governmental officials, reformers, and intellectuals.

Historians of industrialization in England, France, Germany, and North America have produced a very rich body of sometimes controversial studies about child factory labor.[4] They range from accounts that portray child factory labor as the worst evil spawned by nineteenth-century capitalist modernization and view children as its victims to studies that emphasize the Industrial Revolution's positive implications for children's lives.⁵ Perhaps the grimmest picture in modern scholarship of child abuses during industrialization appears in James Walvin's study of childhood in England. According to Walvin, "children were beaten awake, kept awake by beating and, at the end of the day, fell asleep, too exhausted to eat."⁶ In his seminal *Making of the English Working Class,* E. P. Thompson claims that "exploitation of little children ... was one of the most shameful events in [British] history."⁷

In contrast, a few historians offer more favorable assessments of child labor during industrialization.⁸ They maintain that working conditions for children during the Industrial Revolution were no worse and in many cases even better than those before industrialization or those that existed in the countryside. Clark Nardinelli, for instance, suggests that the exploitation of children did not originate in the Industrial Revolution but in the countryside. Indeed, according to Nardinelli, the new job options created by industrialization and the competitive labor market offered children opportunities to escape the even heavier exploitation at home in cottage industries or in agriculture. "Industrialization," Nardinelli writes, "far from being the source of the enslavement of children, was the source of their liberation."⁹ Nevertheless, most recent studies of child labor find Nardinelli's hypothesis questionable and objectionable. They concur in the older view and offer pessimistic evaluations of the Industrial Revolution's impact on child labor. For example, Nardinelli's argument has been questioned by economic historians from Cambridge University who have insisted that the Industrial Revolution indeed led to the harsh exploitation of child workers.¹⁰

The employment of children in late nineteenth-century Russian factories, an issue no less compelling than in other industrializing countries of the time, remains largely unexplored. Despite the wealth of literature on the workers' movement in general, only a few historians have addressed child factory labor. Merely descriptive and empirical, late imperial studies of child labor explored the issue without any analytical or methodological framework. Their authors tended to replicate large citations from published and unpublished primary sources. Among several late imperial studies of child factory labor, E. N. Andreev's book stands out as the most significant and coherent publication, although it too is largely a collection of unprocessed primary sources. Most, if not all, late imperial scholars were highly critical of children's industrial employment, which they portrayed as morally unacceptable and even outrageous in its consequences.[11] V. I. Gessen's two early Soviet-era monographs (both appeared in 1927), with all the limitations of the period's priorities, agendas, and methodologies, remain to this day the only the major Russian-language investigations of the topic.[12] Highly critical of capitalism, Gessen emphasized the harsh exploitation of children in imperial-era industries and alleged a general lack of state concern for children's welfare. The harsh exploitation certainly occurred, but as this study will show, the question of state concern is much more complicated.

Although some English-language histories of labor in Russia mention the issue of children's industrial employment, more often than not in passing, no books or articles have appeared that subject this important aspect of industrial labor to scrutiny in its own right. In his studies, Reginald E. Zelnik notes the persistence of child labor in imperial Russia's factories. His *Labor and Society in Tsarist Russia* outlines the tsarist government's early legislative efforts to constrain children's employment in industries, and his *Law and Disorder on the Narova River,* which analyzes the 1872 Kreenholm strike, provides an account of conditions for working children at the Kreenholm cotton mill.[13] Michael Melancon's *Lena Goldfields' Massacre* provides valuable data about underage gold-mining workers in Siberia during the late nineteenth and early twentieth centuries.[14] Aside from these studies, which serve to introduce the question, the child industrial labor issue remains a virtual blank page in English-language historiography of Imperial Russia.

This study attempts to fill that page. It investigates child industrial labor in Russia from the late eighteenth century until the outbreak of the 1917 revolution and addresses two main questions. First, in view of the reality of widespread and traditional use in the countryside, what impact did industrialization have on child labor? Second, what did child industrial labor signify in economic and social terms? Tracing the origins, extent, and dynamics of child labor, as well as the social background of employed children, the study examines the causes of child labor during industrialization. It examines child laborers' workday, wages, and working conditions and analyzes the malign impact factory labor had on their health. It also draws attention to how the harsh realities of child industrial labor influenced contemporary attitudes toward and sparked debates about the issue. It shows how these debates affected tsarist social legislation and, finally, evaluates the legislation's effectiveness. In more general terms, this book explores imperial Russia's labor and economic history and in doing so opens up new perspectives for comprehending late tsarist society.

One of this study's major hypotheses is that during the second half of the nineteenth century, the widespread, intensive industrial employment of children, with resulting exploitation and decline of health, produced a sharp transformation of attitudes about child labor, from initial broad acceptance to condemnation. Originally popularly accepted as an appropriate means of apprenticing children, child factory labor in fact had a deleterious effect on children's life and health. As awareness of this harsh reality grew, increasing state and public concern about working children helped form new approaches to the issue that resulted in legislative regulation of children's employment, education, and welfare. All these developments provided an important foundation for general social legislation in Russia during the late nineteenth and early twentieth centuries, a broad topic that requires much more attention than it has received.

The data, analysis, and interpretation that comprise this study constitute a social history of child industrial labor. At the same time, it is rather more than that. The early sections of this book discuss Russian society's attitudes—especially of the peasantry but also of educated, entrepreneurial, and administrative elements—toward childhood and the roles of children in the household and surrounding work areas. Later sections trace these attitudes into Russia's burgeoning factory and urban environment. In Russian historiography, the whole question of childhood in imperial times, it is worth mentioning, has hardly been raised, much less exhausted, with the exception of aspects of education. Soviet childhood has been the beneficiary of wider scholarly attention. Even so, precisely because of the absence of commentaries about childhood during the imperial era, some misconceptions have arisen on the issue. For example, on the basis of modernist interpretations, Catriona Kelly believes that the idea of childhood began to receive "unprecedented attention" in public and state discourse from the 1890s.[15] Perhaps awareness of childhood increased as time went by, but my findings suggest that, although historians have been largely unaware of this, society had a clear vision of childhood and its problems long before the end of the century, as reflected in legislative debates about child labor laws, education, and welfare. This study attempts to provide a context for studies that cover a later period. After all, the very concept of childhood acquired its basic features during the entire late imperial era.

The analysis offered here of state and society's discursive responses to a growing awareness of the threat that factory labor posed to children's lives and health constitutes an exploration into the realm of Russia's civil society. The history of the legislation that resulted from this state-society interaction is as yet either unknown or little noted, as any excursus through existing historical literature reveals. Perhaps of equal significance is the way that the legislation originated—in a distinctly interactive process among state officials at virtually all levels and society, as both officialdom and society responded not only to perceived violations of humane norms but to workers' objections and demands. Through strikes and other forms of protest, workers made known their plight and gave testimony to inquiring officials. All of these aspects raise many questions about scholars' approach to late tsarism and its allegedly purely autocratic habits.

For scholars of child labor, as for students of labor history in general, sources and their reliability remain a crucial problem. Therefore, whenever possible I have tried to integrate and balance all available evidence. The book utilizes a wide array of surviving primary documents, as well as published sources, governmental materials, laws, and secondary studies. It incorporates data from many previously unpublished archival documents, published memoirs, and the era's periodical publications.

Published sources include government reports and reports of factory inspectors, health records, labor statistics, business reports, and journalistic accounts. Most of the sources used for this study do not come directly from children themselves, who all too often left no contemporary record of their outlook and experiences. Consequently, this is a study of child workers' experiences as seen through the eyes of adult contemporaries. It is also unavoidably affected by the morals, perceptions, and biases of today's world, including this author. What in fact the working children of that day thought about themselves and their labor in the factories is an almost closed book. Perhaps they were not as miserable as we might assume. Doubtless, in the way of children everywhere, they were often happy and playful despite all the burdens imposed upon them. They hardly saw themselves as exploited victims of advancing capitalism, but rather as young persons helping their families, or achieving independence and self-reliance, or attaining a working trade for the future, or even all of the above. The following chapters strive to provide at least a glimpse into the realities of working children's lives.

The study begins with an exploration of traditional perceptions of children and childhood and analyzes child labor in the countryside—in agriculture, in domestic industries, and in state industries. Traditionally, the use of children in productive labor had been widely accepted, particularly among the lower social classes. The initiation of children into some kind of work was viewed as a form of upbringing and education aimed at preparing them for adult responsibilities. The extent of child labor depended on the economic condition and size of the family. Most families in preindustrial Russia depended for economic survival on the labor input of all family members with the exception of very little children and those unable to work. At this point, let me emphasize that the economic conditions of individual families influenced only the intensity of child labor, not the fact of children's engagement in productive activities. The types of labor assignments for children differed in accordance with the individual child's gender and age.

Initially, the state concurred in the popular view that children's involvement in productive labor served as an education and apprenticeship for adult occupations. Long before the nineteenth century, the apprenticeship of children had been an established and entirely legal practice. With the purpose of helping children "learn a profession," the government sanctioned sending thousands of urban and rural children to state and manorial factories. Reality, however, often differed from intentions. Alongside apprenticeship or even instead of it, many entrepreneurs employed children for regular work, over long hours and even at night. The government undertook some fragmentary measures limited to certain industries and factories in its first timid attempts to cope with the abuses of child labor. The most important legislative act was the 1845 law that prohibited nighttime work for children under the age of twelve. For the most part, however, the early laws lacked uniformity and were quite specific: they aimed only at concrete situations. Thus, by the mid-nineteenth century, the starting point of heavy Russian industrialization, child productive labor had been a widespread traditional and legalized practice, welcomed by most social classes and supported by state laws.

During the late imperial period, Russia witnessed rapid industrialization. The accelerating tempo of the capitalist economy created a massive demand for semiskilled and unskilled labor. This was complemented by rapid population growth and changes in the rural economy after the 1861 reform, both of which led millions of rural residents to seek factory work. Because of the broad popular acceptance of child labor as a means of education and apprenticeship and because of the dependence of most families on the labor of all family members, parents were willing to send their offspring to new factories when the opportunity arose. Simultaneously, manufacturers viewed children as more adaptable than adults to the new factory regime (work hours and discipline) and more capable of learning to work with new machinery and technology. The conjunction of these factors made children an important source of industrial labor. With economic expansion, the absolute number of children employed in factories grew rapidly, although their proportion to adult workers remained stable. As in other industrializing countries, most children in Russia worked in the textile industry, in particular in cotton processing.

The exhausting industrial environment and long work hours had a sharply negative impact on the health of working children. In fact, factory employment led to their outright physical decline. Unlike labor in traditional agriculture and cottage industries, where work was usually conducted under parental supervision, labor in the new mechanized factories subjected children to the rapid pace of machinery and exposed them to moving belts, shifting parts, intense heat and noise, and hazardous conditions associated with dust and the use of toxic chemicals. In addition to general illnesses caused by the new industrial environment, children were quite prone to work-related injuries. The number of such injuries heavily exceeded the incidence among adult workers.

The increasing ill health among factory children and its potential consequences aroused concern among many statesmen and public activists. There was intense public debate about child labor, which often resulted in legislative proposals to regulate child labor. The appeal for child labor protection laws initiated by state and local bureaucrats produced an important discussion of industrial labor among state officials, industrialists, academicians, and reformers. During the early 1860s, the government organized various commissions to inspect and review existing factory legislation in order to work out new provisions. Ultimately, these provisions came together in a first legislative proposal. In 1860–61 this proposal went to provincial governments and industrialists' associations for review and discussion. The ongoing discussion about child labor reform broadened lawmakers' perceptions of the entire phenomenon of child labor. As time went by, the usually unrealized legislative approaches became more and more complex. For instance, later initiatives addressed such issues as children's education and welfare, which had been entirely absent from previous versions. Debates about children's employment in industry during the 1860s and 1870s produced little significant legislation. Nevertheless, these discussions lay an important conceptual foundation for laws of a decade or so later that aimed at regulating child labor and promoting children's education and welfare. Equally important, the debates facilitated the actual introduction of these laws.

In addition, by giving publicity to child-related questions, the debates about child factory labor opened up a new issue in Russian public commentary—the issue of childhood. Childhood became a subject of public discussion and began to receive increasing attention from state institutions; legislators; public groups, including philanthropic societies; and individuals. From the early 1870s, the discussion of childhood became increasingly politicized and was used by various interest groups for their own political or economic agendas.[16]

Starting with the introduction of the 1882 law, the state progressively restricted children's employment in industry and introduced compulsory schooling for working children. It is not commonly realized that the late imperial decades witnessed an unprecedented degree of children's participation in social and political activities such as labor protest and strikes. This involvement in many cases led to their entry into radical political movements.

This study of child labor contributes to an already existing rich scholarship on the labor movement and workers in Russia. Generally, most historians who have paid special attention to the issues of factory labor and workers have been of a leftist inclination.[17] Consequently, Marxist and similar methodological approaches have dominated the scholarship: social categories other than class and the relations among classes have been absent. Marxist-oriented scholarship has made enormous contributions but has also omitted important aspects of the labor experience. In recent decades, some studies have gone beyond class relations by exploring labor issues in relationship to gender, scholars reminding us that the abstract category of class in fact represents individual men and women.[18] In this new approach, gender relations and politics are as important as class for understanding labor issues. Taking gender-oriented labor history as its inspiration, this study offers age as a category of analysis and explores working children as a social group that has its own important distinctive cultural and socio-psychological features.

An understanding of the experience of factory children can deepen our knowledge of Russia's labor history, its industrialization, and its evolving legislative approaches, thereby shedding new light on governance in late imperial Russia. This approach suggests a new interpretation of child productive labor, showing how the transition from the preindustrial to the industrial economy influenced the practice and the extent of child labor. It also contributes to a new understanding of the "preindustrial" concept of childhood. In addition, it suggests a new understanding of late imperial Russian state and society and the relations between them, especially as regards society's participation in the processes of imperial lawmaking. This study proposes a new way of viewing and interpreting the developmental dynamic of Russian society and shows how this dynamic influenced the late imperial Russian state.

Finally, this book offers an excursus into the history of a problem that still plagues many societies today. According to the International Labor Organization, 218 million children worked in industries and agriculture in 2004. About 126 million of them were employed in hazardous work. This study will, I hope, provide some historical perspective on child labor and state efforts to eliminate it. The lessons drawn from history and the insights gained may prove useful for contemporary policy makers in developing their strategies to cure this social disease.

1 ❋ Origins of Child Industrial Labor

CHILD LABOR IN RUSSIA was hardly a product of late nineteenth-century industrialization. Children's engagement in productive activities had existed well before modernized factories began to appear in Russia's primarily rural landscape. From time immemorial, children had worked in agriculture, as well as in cottage industries and all other types of domestic manufacturing. In addition, Russian children worked in manorial and state factories and mines. The principal goal of having children engage in productive labor during these earlier times seems to have been less economic than educational. Child labor had been a widely accepted and common practice, aimed at teaching children adult occupations and thus preparing them for adult life.

Eighteenth-century travelers to Russia often commented on child apprenticeship in their descriptions. When the German geographer Johann Georg Gmelin visited the Demidov Nizhne-Tagil'sk metallurgical works of western Siberia in 1742, he noted with some admiration that "in the wire shop children from ten to fifteen years old performed most jobs and did them not worse than adult [workers]." In the Nev'iansk mill, the geographer observed how seven- and eight-year-old boys made copper cups and various kitchen wares and "were rewarded according to their work." Gmelin claimed that in some workshops, the number of children even exceeded the number of adult workers.[1] Another famous German traveler, Peter S. Pallas, who visited the Urals' mines and metallurgical works (in western Siberia) during the 1770s, wrote that he was "highly delighted to see that young ten- and twelve-year-old children work in the blacksmith shop and receive a salary" on a par with adult workers. Pallas pointed out that the number of children employed in the works reached the thousands.[2] As troubling as they may appear to modern sensibilities, these almost adoring portrayals of the phenomenon of working children reflect widespread contemporary perceptions of child productive labor.

How did child productive labor emerge? What was its nature and extent before industrialization? To track the origins of this phenomenon, it is necessary to explore the role of child labor in the countryside and children's employment in state and manorial enterprises, along with popular notions of childhood and the influence of these notions on child labor and state policies regarding children.

Child Labor in the Countryside

Most scholars of the history of the family and childhood suggest that child labor was common throughout the history of the household, especially when the household was the basic unit of production. Customarily, children worked in agriculture and in cottage industries. Poor economic conditions usually receive emphasis as the major cause of child labor. Some scholars also point out that the use of children in production reflected traditional beliefs about and practices of child rearing and education.[3] All of this was true of Russia. In most social strata, particularly in peasant families and the lower urban orders, initiation of children into some kind of productive labor "appropriate to their strength and ability" was perceived as a form of education and apprenticeship and aimed at preparing children for adult responsibilities. Ethnographers note that in peasant families, teaching household activities and agricultural occupations was considered the most essential duty in the upbringing and education of children. The surviving evidence suggests that in cases when foster parents reported to the village commune about fulfillment of their parental duties, they usually underscored their efforts to teach the children they had adopted all common household and agricultural occupations. Peasants believed that "if a child is not initiated into productive work from an early age, it would hardly develop the ability for work in the future."[4] Here I would like to emphasize that child labor as form of preparation for adult life, a practice common before industrialization, is also perhaps the most important characteristic of children's regular employment in industry after the beginning of industrialization.

Child labor's crucial educational aspect aside, its widespread acceptance also signifies the extent to which most families of preindustrial Russia depended for their economic functioning on labor contributions from all family members, including children and elders. Here, however, is where this account differs from scholarly views that emphasize the impoverishment of peasant families as the primary cause of child labor in the countryside.[5] This commonly held view hardly seems adequate. Rather than poverty, the origins and development of the local peasant economy within the context of the family influenced the use of children in production. Obviously, an individual family's economic conditions might affect the degree of children's participation in productive labor. These conditions were, however, hardly its consequential cause. Because the family was the basic unit of production in the preindustrial economy, the nature of the peasant household and the conditions of its maintenance required the labor input of all family members, with the exception of very small children, usually under the age of five, and very old people. Thus, child labor was essential for the economy of every peasant household in preindustrial Russia, regardless of its economic conditions. The upbringing and education of children went side by side with the real productive economic activity of the peasant family.[6] Even so, apprenticeship, rather than potential contribution to family income, was perhaps the most important cause of child labor in the countryside.

An old peasant custom of calling juveniles by names according to their labor task illustrates the wide popular acceptance of child productive labor. For example, boys between ages seven and ten who helped plow or harrow were called *pakholki, paorki,* or *boronovolki* (plowboys or harrow boys); those who helped pasture animals were called *pastushki* (herd boys). Girls of the same age were called variously *nian'ka* (nanny girls), *pestun'ia* (mentor girls), *kazachikha* (maids who worked as domestic servants in other families), and so on—all names that reflected occupational activities. "Our plowboy," "our herd boy," and "our nanny girl" were habitual terms parents used to address their children.[7]

Did peasants distinguish childhood from other stages of life? One of the principal questions in childhood studies centers on the origins and development of childhood as a concept. Beginning with Philippe Aries, scholars have widely viewed childhood as a cultural invention of modern times. Exploring European arts, Aries asserts that premodern Europe "did not know childhood [and] did not attempt to portray it.... The idea of childhood did not exist."[8] Following this "modernist" approach, many scholars have argued that medieval and early modern society did not see children as persons in a unique and separate stage of life, but rather perceived them as "miniature," underage adults. Accordingly, this suggests that peasants did not separate children from persons of other ages. This conclusion seems to be at odds with some recent studies and with the findings of Russian-language ethnographers and anthropologists. These scholars suggest that peasants, in general, distinguished three major periods of the life cycle—childhood, adulthood, and old age—with complex subdivisions and stages within each period.[9] These divisions not only rested upon popular attitudes about human biology but were also embedded within a broad range of cultural assumptions and social roles. According to I. I. Shangina, in the countryside, the criteria for transitions from childhood to adulthood and from adulthood to old age were relative and depended on the individual's physiological condition and readiness to undertake one or another responsibility.[10]

In general, these divisions usually corresponded with the individual's ability to work and clearly reflected peasant practices of distributing labor duties among family members. Labor duties in peasant families were carefully defined according to the age, gender, and physical abilities of family members. Full working age depended on life span and normally ranged from about seventeen to sixty-five, a group that included roughly 60 to 64 percent of the peasant population.[11] Very small children, under the age of five or six, and people over sixty-five usually did not work. Children between ages eight and fourteen were considered "half workers of little strength" (*polurabochie maloi pomoshchi*), whereas juveniles between fourteen and sixteen years of age were "half workers of greater strength" (*polurabochie bol'shei sily*). In the countryside, the age of peasants when they received "full labor duty" (*tiaglo*) varied from province to province. On average, starting from the age of seventeen or eighteen, peasants carried full labor duty until somewhere between age sixty-one and sixty-five. The full state poll tax was assessed on adult peasants, usually beginning at age eighteen. Juveniles aged fifteen to seventeen years were subjected to half labor duty. In impoverished families, or in families in which one of the adults was absent or deceased, children fulfilled all adult responsibilities at an earlier age.[12] Thus, while child labor in general resulted from the described cultural factors, in some specific circumstances, it was related to economic causes.

Various studies of rural youth indicate that children under age fifteen constituted a significant portion of the peasant population, about one-third. According to Baklanova's findings, children under age five accounted for about 14.0 percent of the peasant population of northern Russia, children between six and ten constituted about 11.0 percent, and those between eleven and fifteen made up about 9.0 percent.[13] During the nineteenth century, children age seven and younger accounted for about 17.5 percent of the population of European Russia. (In 1858, the population of European Russia was 59.2 million, of which about 49 million were peasants.)[14] Infant and child mortality rates, however, were high. During the nineteenth century, only about 50.0 percent of children survived to age ten.[15] Such a high mortality rate among infants and young children was typical of most of preindustrial Europe. Anthropologists tell us that the high death rates among infants resulted from the common practice of sending babies away to wet nurses hired by the family. For example, in mid-nineteenth-century France, about 25.0 percent of such infants died before the age of one, and only 50.0 percent survived to age five. Similar figures come from other European countries. Scholars also point out that high infant mortality resulted from alleged widespread neglect of and indifference toward children in peasant families.[16]

In contrast, Russian-language anthropologists maintain that the high death rate among children resulted not from neglect or indifference but from poor living conditions and inadequate medical knowledge. Russian peasant families in fact highly valued their children precisely because of the high mortality rate among them and because of their potential significance as future household and agricultural laborers. One popular peasant saying holds, "Our own harrow-boy [*boronovolok*] is much more valuable than anyone else." The adult population in village communes provided children with love and care, as well as tolerating some of their mischief. Children were considered young and silly and therefore were easily forgiven for pranks. This, however, did not exclude punishment applied within the individual family. When punishment occurred, parents were careful not to cause serious physical harm to their children. A former serf from Yaroslavl' province of central Russia, Savva Purlevskii, recalled that when he was a child in the early nineteenth century, he was beaten by his father "only on rare occasions," because, as he explained, his parents were concerned about his health. His "grandmother would not let anyone beat [him], because [he] was the only child they had."[17] The value of children as future laborers helps explain the low rates of infanticide and child abandonment in Russia.[18]

The high mortality rate among infants and young children affected the number of children in peasant families. Table 1.1 shows the number of children in peasant families in 1717 in the Kubenskii region of Vologda province (northern Russia). The figures show that about 47.0 percent of surveyed peasant families had two or three children, whereas 22.4 percent had one child. It is worth noting that about 14.0 percent of peasant families had no children at all. It is not clear, however, whether these families never had any children or had none at the time of census. Nevertheless, an average family had two or three children. In Russia, extended two-generation families with two adult male and two adult female members predominated.[19]

As mentioned, evidence suggests that Russian peasants distinguished childhood as a unique stage of life. Researchers of popular culture have noted that peasants considered childhood to last from the moment of "coming into this world" until about age fifteen or seventeen. Depend-

1.1 Number of Children in Peasant Families in the Kubenskii Region of Vologda Province in 1717

Number of Children	Peasant Families	
	Number	Percentage
1	238	22.4
2	292	27.4
3	208	19.5
4	84	7.9
5	59	4.6
6	22	2.1
7	10	0.9
8	5	0.5
9	3	0.3
10	1	0.1
11	1	0.3
No children	150	14.2
Total	1,064	100

ing on locality, childhood's "upper limit" ranged from thirteen to nineteen.[20] Peasants considered infants and very young children, from birth to five or six, as neutral—without gender. Collective names for children of this age thus did not reflect their gender, although personal names were given according to a child's biological sex. Regardless of sex, both male and female infants were variously called *ditia, rebenok, mladen', and momzik,* all of which can be translated as "baby" or "child." Small children were also called *kuviaka* or *kuvatka* (those who cry), *sligoza* (those who drool), *popolza* (those who crawl), and so on, depending on locality.[21] These names do not reflect a child's biological sex but rather suggest either age (*ditia* and *rebenok*) or behavior associated with a particular age, such as crying, crawling, and so on.

The clothing of very young children also did not distinguish their biological sex. Peasant children of both sexes usually wore long linen shirts until five or six years of age. Young boys, until that age, normally did not wear pants. In most peasant families, children's clothing was produced from old, worn-out adult clothes and was passed from older children to younger ones.[22]

Peasants believed that a child's biological sex—or, as they called it, "natural" sex—did not automatically translate into the proper social behavior normally attributed to that biological sex. Parents utilized various customs and activities associated with magic and popular religion in order to "fix" a child's biological sex. In other words, peasants carried out certain activities to encourage the development of their children in a way considered appropriate to their biological sex.

This process of "fixing" started early, from the day of birth. For example, in many northern provinces of Russia, parents tied the umbilical cords of newborn baby boys under an object that they associated with traditional male occupations, such as a hammer or an ax, whereas they tied those of baby girls over objects associated with female occupations, such as spindles, yarn, and so on. These objects were related not only to male or female spheres of activity but also to the occupations that parents desired for their children's future. For example, depending on the parents' desires, a daughter's omphalos was cut over a spindle or a thread, whereas a son's was cut upon a hammer, an ax, a form for making peasant bast shoes, and so on.[23]

Some practices of "fixing" biological sex involved magical handling of the child's placenta. For example, in Orel province in south-central Russia, a mother would take a piece of her baby's placenta and put it in a place or upon an object she associated with the child's desired future occupation. In Vologda province of northern Russia, the father would hang the placenta of his baby son in the stables, while saying, "The child grows up with the horse."[24] The different places chosen for boys and girls clearly indicate that peasants distinguished male and female spheres of productive activity. By magically associating children with one or another sphere, parents tried to encourage behavior appropriate to the child's biological sex.[25] This finding perhaps supports the idea held by many gender theorists that gender is a socially constructed category.

In general, with a few exceptions, the initiation of children into agricultural, household, and other productive labor started early, usually from age five or six, and involved very simple tasks. As children grew up and acquired greater physical strength, parents gradually assigned them more complicated and serious tasks. The initiation process in some cases was accompanied by additional ritualistic activities and rites. Anthropologists believe that the latter symbolized the transition from childhood to adolescence.[26] For example, in Smolensk province in western Russia, a young girl, at age five or six, was assigned to spin a single thread for the first time in her life. Then the thread was burned, and the girl was supposed to consume its ashes with water and bread. This ritual was accompanied by a saying: "Eat and you will become a good spinner."[27] In other areas of Russia, boys and girls around age five to seven began to wear pants and skirts, modes of dress that also symbolized their transition to a new stage of life.[28]

Thus, the transition to adulthood began around age five with the symbolic introduction of children into productive activities and continued for the next several years. During these years, children were characterized as "undergrown," "underage" (*podrostkovye*), or juvenile. This observation puts into question another modernist hypothesis—namely, that the period from birth to six years of age held a child's full transition to adulthood. In this view, peasant children began to carry out all adult responsibilities from age six.[29] Findings from the Russian countryside, however, suggest that the transition to adulthood, rather than being completed by age six, started around age five and continued for years thereafter.

In order to facilitate the initiation of very young children into one or another productive activity, parents developed various treats and rewards or assigned simple labor tasks in the form of play. Many games and recreational activities children engaged in between the ages of six and fourteen imitated adult occupational and social activities. For example, in some areas, peasant children played *konople,* a game that mimicked certain labor tasks necessary to hemp cultivation.[30] One investigator of children's recreational activities remarks that "a game was a particular way of preparing children for adult life."³¹ Sometimes, in order to wake up small children early in the morning to move hay or plow, parents used some kind of special treat. For example, an ethnographer in 1856 recorded peasants in the Altai region (western Siberia) putting baked eggs by a sleeping boy, saying, "Wake up, little Peter [Petushok]. The hen has already laid two little eggs by your head for you."(In the Russian language, *petushok* also refers to a young rooster.)³²

In general, children performed various types of work according to their gender, strength, and ability. Boys were usually launched into traditional activities of adult male peasants. Young sons were expected to help their fathers sow and thresh and cart manure to the fields. For example, in the Narymsk province of western Siberia, boys around age five or six began to assist adult peasants in manuring soil. The most widespread communal function for six- or seven-year-old boys was herding animals. Boys who engaged in herding were called *podpasok* or *pastushok* (herd boy, shepherd boy, cowboy). At about the same age, boys in many provinces also began to learn how to ride on horseback. In most cases, young boys worked under the supervision of their fathers or older male children.³³

As boys grew older and gained more physical strength and ability, parents gave them greater responsibilities and assigned them more complicated tasks. At the age of seven, eight, or nine, boys began to help adult peasants with land cultivation. In the Shadrinsk district of western Siberia, boys of this age and occupation were called *pakholki* and *boronovolki* (plow boys, harrow boys). Their work involved leading horses during plowing and harrowing. From the age of nine or ten, boys began to carry out various other activities: accompanying the cows to water, feeding animals, carting manure, harrowing, helping adults in plowing and harvesting, and carrying provisions for adult males who worked away from the village (in local forests or on nearby rivers, ponds, and so on).³⁴ From age thirteen to fourteen, the male peasant was supposed to work with the scythe, sickle, thresher, and ax and began to learn how to work with the plough. At age fifteen, the son became a "full assistant" (*polnyi pomoshchnik*) of his father and could replace him in case of absence or sickness.³⁵

Girls helped their mothers maintain the household; cared for the younger children; and carried out all the agricultural responsibilities of adult female peasants: raking, strewing, reaping, binding sheaves, gleaning, and so on. Depending on the province, girls also learned various crafts and cottage industries, which in Russia were predominantly female endeavors. Girls' occupations were usually within the household or the local community, whereas boys' activities were inside as well as outside the village. Nevertheless, the occupational roles of boys and girls were sometimes interchangeable. In families without male children, girls helped with agricultural tasks normally performed by boys; in families with no female children, boys helped with female work.[36]

Regional economic variations also determined the character of children's occupations. In areas where agriculture predominated—southern, western, and central agricultural areas; the Volga provinces; and Siberia—children performed mostly agricultural tasks. During the nongrowing season in agricultural areas, children engaged in various domestic industries and types of work not associated with farming. While girls usually stayed at home helping female peasants, boys often migrated with fathers and worked away from the village. By the late nineteenth century, with the growth of industry, the seasonal migration of rural children to industrial centers had increased significantly.[37]

In addition to farming, boys helped parents with hunting and fishing in areas where these activities were part of the local economy. From the age of eight or nine, boys learned how to use the bow and how to set up nets on lakes and ponds for catching wildlife. Initiation into fishing and hunting started as play and gradually took more realistic forms. Finally, as they grew older, boys were invited to engage in real hunting and fishing, beginning with the simplest and easiest assignments and then moving on to more complicated and difficult ones.[38]

In regions where the local economy was mixed or predominantly nonagricultural, children engaged in cottage industries and crafts. In central nonagricultural provinces, parents taught children textile making and other crafts characteristic of the local economy.[39] Girls engaged in various cottage industries, whereas boys were initiated into commercial activities or worked outside the village. For example, Savva Purlevskii, the serf mentioned above, was from Yaroslavl', a nonagricultural province in central Russia. He recalled in his memoirs that from an early age he engaged in petty trade. At the age of eleven (in 1811), Purlevskii lost his father and from that time on had to earn his living. At the age of eleven, as he put it, "the laboring part of [his] life began." His mother wove linen canvas, and he sold it in local and regional markets. In addition to selling his mother's canvas, Purlevskii also bought flax and other locally produced peasant goods and transported them to Moscow or local markets where he sold them. As he grew up, he traveled for longer distances.[40] In these respects, the young Purlevskii's activities were typical of Russia's central nonagricultural provinces, where peasant trading activities were widespread. In Vladimir province, famous for its nonagricultural economy, the *ofeni*—local male peasants who engaged in commerce—took their children to Ukraine, Volga, Siberia, and everywhere else in the empire where they engaged in trade.[41]

In areas with mixed economies, girls usually remained at home and learned various crafts and trades. In Russia, peasant cottage industry was virtually a women's sphere. According to a 1787 observer, "Women of [the Nikitskii district of Moscow Province], as is usual everywhere [in central Russia], spin flax and wool and weave canvas and cloth for household use and for sale." This observer recorded similar activities among women in other nonagricultural provinces of central Russia.[42]

Most evidence suggests that Russian peasant children made the transition to adulthood, at least in terms of occupation, at about fifteen years of age. By this point, peasant boys and girls were supposed to have learned how to accomplish a certain number of occupational tasks. Those who could not learn how to do work appropriate to their age were subjected to mockery. For instance, a girl who could not learn how to spin by a certain age was called a "no spinner" (*nepriakha*); if a girl could not weave cloth by age fifteen, she was called a "no weaver" (*netkakha*). Boys who had not learned how to make bast shoes were called "shoeless" (*bezlapotnik*). One contemporary observer noted that male peasants who could not make peasant shoes were not respected by fellow villagers and were generally viewed as "losers."[43]

In addition to a given child's age, gender, strength, and ability, village children's occupations and the extent of their engagement in productive labor depended on economic and demographic factors. The economic status of the family, its size, and the number of adult workers were perhaps the most important factors. Various studies illustrate that in families with no adult male workers, all responsibilities fell upon women and children. This was especially true for nuclear families where male members were in the military or deceased. Children's labor input in these families was greater, their responsibilities more extensive, than in families with two or more adult workers. According to T. A. Bernshtam, "The scarcity of men's hands in a family led to its economic decline, whereas the absence of men in a nuclear family led to poverty." Labor pressure on older children was also heavier in nuclear families with small dependent children. Most families in preindustrial Russia, however, were "traditional extended," including two or more adult males. Studies by Russian-language anthropologists suggest that living conditions for children in these families were better than in Russian nuclear families.[44]

Historians of child labor in preindustrial Europe emphasize the poor living conditions of most peasant children before the Industrial Revolution. Many children had to start their laboring lives as early as four years of age and therefore had, in the words of Mary Lynn McDouglass, a "short childhood."[45] Scholars suggest that in European nuclear families, children were often treated with indifference and neglect. About 25 percent of children died before age one and 50 percent before age five.[46] One study of Manchester workers in 1842, for instance, asserts that "more than 57 percent" of the children of the city's "laboring classes" died before age five.[47] Relying on Russian-language scholars, who have produced a rich body of studies on the family, one may surmise that prior to industrialization, the conditions of children in Russia's lower-class extended families were better than those of most of their European counterparts, who lived in small nuclear families. Nevertheless, the average child-mortality rate in Russia, across all social segments, was relatively higher than in Western Europe because of various ecological, economic, and social factors.

Was there exploitation of child labor in the countryside before the Industrial Revolution? This question provokes highly contradictory responses from historians of child labor. Clearly, the extent of exploitation of children in the countryside depended on the specific time and place in which the children lived. As scholars suggest, in small nuclear families and in families with no adult males, child labor might prove to be more economically significant, and children's labor burden therefore heavier, than in extended families. Furthermore, capitalization of the rural economy in Russia during the nineteenth century, as the market economy intensified, may also have led to an increasing labor burden on children in individual families (see chapter 2).

Nevertheless, it is probably safe to suggest that the purpose and nature of children's involvement in productive activities in the Russian countryside differed from what might be expected. Parents did not put their young children to productive work for the sake of profit. Historians of child labor suggest, rather, that children's agricultural productivity was usually low, greatly lagging behind their consumption until age thirteen or fifteen. The same was true for the nonagricultural sector of the rural economy.[48] The only possible conclusion is that parents in the countryside engaged their children in productive work primarily for the purpose of teaching and apprenticing them. As we have seen, the introduction of children into productive labor was generally a gradual process that lasted for several years, until the child grew up and was finally assigned an adult workload. Moreover, children were given tasks according to their gender, physical strength, and abilities, and they worked under the supervision of their parents or other adult family members. To reiterate, the ultimate goal of child productive activities in the countryside was to prepare children for adult life—to help them become fully functioning members of an individual family and community. Perhaps we should cease using the term *child labor* to describe young children's involvement in production in the countryside and instead call this practice what it was for peasant families—a form of schooling, education, or apprenticeship. These pragmatic arrangements taught children what they would be expected to do when they became adults.

During the second half of the nineteenth century, as industrialization intensified in Russia, production switched from the family and individual household to mechanized factories, where work practices involving child and family labor received wide acceptance. As important a transition as this was, it nonetheless did not mark the beginning of children's involvement in factory labor in Russia.

Child Labor in State and Manorial Factories

Long before industrialization, children, aside from being involved in productive labor in the countryside, also worked in state and manorial industries. A brief description of state and manorial factories will help situate child labor in these enterprises within an appropriate historical context. Manorial (*votchinnye*) and state-owned (*kazennye*) factories, which dominated Russian industry in the seventeenth and eighteenth centuries, relied largely, although not exclusively, upon the labor of paid hereditary serfs. Most of these industries were in reality traditionally organized craft workshops, with low levels of mechanization. From the end of the eighteenth century, the number of hereditary serf workers in these enterprises began to decline, giving way to freely hired contracted labor. The imperial decrees of the late eighteenth and early nineteenth centuries allowed owners or managers of manorial and state factories to free their hereditary serfs. Alongside legislative measures to eliminate bound labor, it also declined because many state enterprises underwent privatization during the first half of the nineteenth century. New, privatized businesses preferred to employ contracted laborers. Subsequently, hereditary serf labor declined steadily throughout the 1840s and 1850s and fully disappeared with the 1861–64 reforms.[49]

As in the countryside, work in these concerns usually served as a way of educating children in a profession or craft. The tsarist state concurred in the popular view of child labor as an education and apprenticeship. State officials believed that children over ten or twelve years of age should engage in one or another kind of productive work, "according to the child's age, gender, and strength" (a formulation that dotted imperial commentaries on the subject). In 1811, a state official inspecting the Krasnosel'skaia state textile mill found it "unacceptable" that the mill workers' sons under age fifteen "did not work at all." His inspection resulted in the issuance of a special Senate decree for this mill that obliged male children of mill workers to obtain an apprenticeship by age twelve. Later, the prominent statesman and public activist Admiral Count N. S. Mordvinov, who from 1810 to 1838 headed several departments of the Imperial State Council, maintained that children of peasants and the lower urban orders "could serve with great usefulness" in the nation's industrial development. Similarly, General Count A. A. Zakrevskii, who from 1848 to 1859 served as the military governor of Moscow, gave child labor a favorable assessment, arguing that employment of children in industries could bring "great benefits for working families."[50] Both officials carried great weight in imperial policy making. Their views exemplify the state's early acceptance of popular perceptions of child labor and its policies regarding working children.[51]

Although archival materials containing social data about state industries' workforces are extremely sparse, this lack is compensated by relatively abundant imperial decrees and surviving legal records. Imperial decrees on this subject were very specific well into the nineteenth century, often relating not only to a particular realm of industry but to individual concerns. State decrees and regulations regarding children suggest that the state not only embraced popular views about children's involvement in productive labor but also accepted popular perceptions of childhood. For example, legal documents and decrees often distinguished among three categories of children based on age: children under age eight, children between eight and twelve, and children between twelve and eighteen. In the text of state decrees, children between eight and twelve were customarily designated "underage" (*maloletki*), whereas those between twelve and eighteen were called juveniles (*podrostki*). For instance, an 1818 regulation of serf workers' maintenance and living conditions in the Ekaterinoslav state mill considered workers' children under eight completely dependent on their parents: their food allowance was given to their parents, whereas children between eight and ten received their own food rations. As they reached age ten, serf children became apprentices and—in addition to or sometimes instead of food allowances—received wages.[52] Legal records from other enterprises also suggest that hereditary serfs' children received food allowances until they reached age eleven or twelve. After that age, they were expected to begin an apprenticeship, for which they received financial rewards.[53] Some state-enterprise regulations provided children under age twelve with small monthly cash allowances.[54]

The state also seemed to adhere to the popular pattern in the matter of initiating children into productive labor. Following the practice of giving children more serious and complicated tasks beginning around age ten or twelve, the state regarded this age as suitable to start an apprenticeship that would last until age sixteen or eighteen. A new 1847 statute regarding the mining industry, for instance, obliged eight-year-old children of serf workers to attend schools maintained by the mines. In two years, after completing a two-year course, they were to become apprentices in the mines or be sent to a higher-level district school. It is interesting to note that children between ten and fifteen, with their parents' agreement, could be assigned "light" ancillary work, "according to the children's age and strength." Those who reached age eighteen became regular mine workers. As noted, expressions such as "according to the children's age and strength" and "work that fitted children's age and ability" appear over and over again in legal documents that address children's apprenticeship and employment.[55]

State attitudes about childhood, which clearly echoed popular notions, influenced state policies toward children. By the end of the eighteenth century, the apprenticeship of children in state and manorial industries was already firmly established and sanctioned by law. The earliest decrees of apprenticeship date back to the reign of Peter the Great, who famously strived to facilitate Russia's economic development and promote industry.[56] With the purpose of helping children "learn a craft" and "gain a professional education," the state sanctioned sending hundreds of urban and rural children, including the inmates of foundling homes, to state and manorial factories. For example, in 1804, the Imperial Senate issued a decree that sent twelve- to fifteen-year-old orphans and poor children from St. Petersburg to the Aleksandrovsk Textile Mill "to learn the textile craft."[57] Many such apprenticeship decrees appeared during the late eighteenth and early nineteenth centuries.

On the one hand, some of these decrees suggest the desire to apprentice the best available youths for what were perceived as important tasks in a particular concern. They stipulated that admission to apprenticeship be carried out on a selective basis and maintained that only those children who displayed "the ability to learn" and "had not shown any [tendency toward] bad behavior" could be accepted as apprentices.[58] On the other hand, other decrees sanctioned sending ten- to fifteen-year-old children who were attending schools but displayed no "capacity to learn" to factories instead.[59] In reality, most early Russian enterprises, which suffered from a constant need for workers, seemed to accept anyone who wanted to become an apprentice. Apprenticeship in productive labor was perceived as the best form of children's upbringing. Such labor cultivated good morals and prepared the young for an appropriate adult life.

Alongside the educational functions of apprenticeship, the state sometimes viewed apprenticing poor children as a means of providing welfare. Apprenticeship was a measure aimed at combating poverty and crime among the lower classes. For example, a 1722 Senate decree stated that children of Moscow and Riazan' who "wander about on the streets begging" were to be sent into apprenticeship in the cities' factories until they had attained their majority.[60] Another decree, from 1744, allowed the apprenticeship of soldiers' children who had lost one or both parents and did not have the means of survival, "so that [they] would not perish." As noted above, the involvement of children from impoverished nuclear families in productive labor was high.[61]

During the late eighteenth and early nineteenth centuries, many owners and managers of manorial and state factories finalized formal agreements with imperial orphanages to provide apprenticeships for their inmates. Factories promised to teach children crafts and industrial skills, as well as provide them with room and board at present and in the future. For example, in 1798, a textile entrepreneur asked the imperial orphanages to transfer some three hundred orphans to his mill as apprentices.⁶² In 1822, S. G. Gesse, the owner of a cotton mill, asked the Emperor's Orphanage to hand over twenty teenagers between the ages of twelve and fourteen for apprenticeship in his mill.⁶³ As suggested by numerous such agreements between factory managers and orphanages, orphanages during this period became a sort of labor supplier for state and manorial industries, forwarding hundreds of their inmates to factories.⁶⁴

In order to find children for their enterprises, employers sometimes traveled around local villages and towns looking for potential recruits. For example, an 1818 account from the Altai region's iron-ore mines and metallurgical works stated that "beginning in the early spring, employers recruited [seven- to twelve-year-old] children [to work] in the mines and mills. Centers of this recruitment were the cities of Zmeinogorsk and Salair, from which children were sent out to various mines and factories [of the region]. In Zmeinogorsk about 500 to 800 boys were recruited each year." The account maintained that during the spring and summer, children engaged in ore sorting and other "easy" tasks, whereas during the winter they were supposed to attend the mines' schools.⁶⁵ In some cases, entrepreneurs especially preferred hiring children. For instance, in the late eighteenth century, a group of owners of Moscow textile mills stated that they had a great need for ten- to fifteen-year-old children and requested that the government provide them with such. These entrepreneurs insisted that without children's labor, certain operations could not be completed, and the whole business would come to a halt.⁶⁶

The government also provided state and manorial factories and mines with the legal basis for using workers' children to labor "according to the children's age, gender, and strength." Numerous imperial decrees allowed state and manorial factories and mines to employ the ten- to twelve-year-old sons of workers in labor "that fit the children's age and physical ability." For example, a statute "on the improvement of the Pavlovskaia Wool and the Ekaterinoslav' Leather mills" stated that all of the mill workers' male children above age ten were supposed to work in these factories and be assigned tasks "appropriate" to their physical abilities.⁶⁷ Workers' daughters, however, could not be required to work without their parents' agreement until they reached age eighteen; after age eighteen, their employment would depend upon their own or their family's desires and needs.[68] According to a 1799 statute on the Urals' mines and metallurgical plants, hereditary workers' sons who reached age twelve and unmarried daughters at age eighteen, with their parents' agreement, could be assigned work, as the laws constantly reiterated, "according to their strength."[69]

Seeking to increase their revenues, landlords who owned hereditary serfs in some cases made agreements with local factories to farm out their indebted serfs, including children. Some landlords had manufacturing establishments on their estates where they employed serf children from indebted families who failed to pay rent. In order to pay off their debts or fulfill other feudal obligations to the landlord, indebted serfs were supposed to work in factories for a certain period of time. In these cases, workers' wages, or substantial portions thereof, went directly to landlords. Available evidence on such agreements indicates that landlords sometimes received from ten to forty-two rubles a year for each child sent to a factory.[70] In 1823 and 1825, the state introduced a series of decrees that banned the forced outfarming of labor, forbade any agreements between landlords and employers regarding serfs, and introduced penalties for transgressors. Forcefully outfarmed serfs could bring lawsuits that sought their freedom from serfdom.[71] Landlords, however, often evaded the law by stating that outfarmed serfs had been sent as apprentices to "learn an industry and receive a professional education."[72]

In addition, the government authorized sending to state industries juveniles who had been accused of committing crimes or engaging in prostitution and those defined by the state as strays or "neglected ones" (*prazdnoshataiushchiesia*). For example, in 1755, a sixteen-year-old peasant boy, Vasilii Fedoseev, was charged with the rape and murder of an eight-year-old girl. The Imperial Senate, which reviewed Fedoseev's case, sentenced him to "harsh punishment with whips" and exiled him to the Nerchinsk mills (in Siberia) for life. Indeed, Siberian industries often used the labor of the children of persons serving life terms at exile or hard labor in Siberia. For example, in 1840, the Iletsk Salt Mines employed 232 children of prisoners who had been sent to the region. An 1849 decree, however, prohibited any further employment of prisoners' children in industries.[73]

In general, according to the laws, the employment and apprenticeship of children, with the exception of children of workers attached to state and manorial factories, was to be carried out with the agreement of the child's parents or, if none existed, with the agreement of local courts or juvenile authorities. Sons of hereditary serfs under age twelve, and daughters under age eighteen, could also not be employed without their parents' consent.[74] Employers, in turn, were required to teach working children a profession; support them "according to their social estate"; provide them with clothing and food allowances; and pay them or their parents a certain amount of money monthly, annually, or upon the completion of apprenticeship. For example, the statute regarding the Pavlovskaia Wool and the Ekaterinoslav' Leather mills obliged the administrations to pay employed children in money and in kind, the latter meaning various cereal crops.[75] After completing the apprentice program, children received twenty-five rubles. Their further work in these mills depended on the mutual agreement of the two parties (children and factory administration).[76]

Although children's employment in most cases required their parents' agreement, contemporaries noted that parents were quite often willing to put their children to work. In such cases, children could earn their own money and contribute to family budgets or themselves pay the poll taxes, which for children from seven to seventeen years old ranged from 0.15 to 1.60 rubles a year.[77] In one surviving petition written in 1803, workers at the Iakovlev Linen Mill complained that their children performed ancillary work and were supposed to receive four kopecks a day. The manager, however, graded children's daily pay according to a scale of three, four, or five kopecks a day, creating discontent among the children's parents and causing complaints. The manager responded that "if workers are dissatisfied with these various rates, let them keep their children at home and support them until they are at least fifteen years of age." In their petition, the workers stated that they had no means of supporting their children other than their children's employment at the mill.[78]

Terms of apprenticeship and employment conditions in state industries were regulated by statutes on state industries and mines. A 1736 statute on state industries, for instance, required employers to teach workers' children skills in industrial trades and crafts so that "they could become competent masters and foremen in the future." According to the Mining Statute of 1806, children of mine workers were paid fifty kopecks a month if they attended mine schools and were not employed in mines and one ruble a month if "they performed work in mines according to their age." In addition, the children received sixteen to twenty kilos (forty to fifty pounds) of flour each month.[79]

In manorial or private businesses, agreements between employers and those responsible for children (parents, guardians, orphanages, and so forth) specified the terms of apprenticeship. Of interest is a surviving formal agreement concluded in 1822 between the entrepreneur S. G. Gesse and the administration of the St. Petersburg Foundling Home, which, as noted earlier, sent a number of its inmates to Gesse's mill. The orphanage was supposed to provide the children with clothing and shoes during their first year in the mill, after which these were to be supplied by the mill itself. The mill was also obliged to furnish the children with "healthy, well-prepared meals" and look after their health and morals. The agreement required Gesse to pay each child from 0.5 to 2.5 rubles a month, depending on the child's behavior and diligence. After the children had gained all the required skills, their monthly wages were to increase to 5.0 or 9.0 rubles a month.[80]

To what extent did state and manorial factories actually use child labor? What work did children in fact perform? It is difficult to estimate the numbers and proportions of children apprenticed or actually employed in state and manorial factories since only fragmentary statistics from single industries and factories are available. Nevertheless, the surviving figures offer a certain insight into the extent of children's employment in individual state and manorial factories. According to a 1737 report, the nobleman Goncharov's Maloiaroslavets Textile Factory (in central Russia) used the labor of 1,719 workers, both contracted and hereditary. Among these individuals, 432 (25.0 percent) were children under age eight, and 211 (12.3

percent) were between nine and fifteen years of age.[81] In 1797, out of the 1,119 workers at the nobleman Osokin's wool mill in Kazan' (in the Volga region), 430 (38.4 percent) were children and teenagers. The Pereiaslavl'-Zalesskii Cotton Mill employed 792 workers, including 183 children (23.1 percent).[82] In 1812, documents show, the state-owned Sestoretsk Armory in St. Petersburg had 195 children with the status of apprentice, at a time when the factory employed 1,244 workers.[83] According to the records of Altai region mines and metallurgical works, these enterprises, by the end of the eighteenth century, employed 19,522 workers, of whom 1,118 (5.7 percent) were children under age thirteen and 603 (3.0 percent) were between thirteen and fifteen.[84] An 1858 description of the Perm' State Copper Works notes that it employed 7,562 workers, of whom 3,377 (44.6 percent) were "underage" children between ten and twelve and 508 (6.7 percent) were juveniles between fifteen and eighteen.[85] Although hardly exhaustive, these statistics suggest the likelihood that most, if not all, state and manorial factories employed children, many heavily dependent on the labor of underage workers.

It would be misleading, however, to assume that all children ascribed to a certain mill actually worked, although many doubtless did. This is particularly true of children recorded in statistics as apprentices or as "children of hereditary serf workers." Such categories reveal very little about the children's real activities in these factories. Moreover, available data often does not specify the children's ages, designating them all as "underage" or "undergrown" (*maloletki*), a broad category that might include very young children, as well as those between twelve and sixteen years of age. Many early Soviet historians of child labor tend to count all children ascribed to an enterprise as "factory workers," assuming that they all engaged in the production process. This tendency prompted scholars to reach perhaps exaggerated conclusions asserting extraordinarily high proportions of working children in state and manorial factories. For example, Vladimir Gessen, the premier Soviet historian of child labor, claimed that "during the first half of the nineteenth century the labor of children between eight and fifteen reached a very large scale ... making up 25 percent of the total number of workers."[86]

Other evidence suggests, however, that by no means did all hereditary serf children juridically attached to a factory actually work there. For example, as we have seen, the sons of hereditary factory workers began their employment or apprenticeship between the ages of ten and twelve, whereas the daughters could not be employed without their parents' agreement until they reached age eighteen. Nonetheless, all male children under ten and female children under eighteen are reflected in factory records as the "children of serf workers." In reality, many children mentioned in the statistics on state and manorial factories—especially very young children—did not work at all. In his highly respected 1923 study of workers, the historian K. A. Pazhitnov argues that "no more than half" of hereditary serfs ascribed to the Altai region mines and metallurgical works actually performed any labor in these enterprises. The same calculation would apply to their children. Furthermore, according to Pazhitnov, children under age eleven were normally employed in state factories only on exceptional occasions. Child labor in these factories, even when it existed, often took on a "sporadic or seasonal" character.[87]

This observation seems to be accurate. As mentioned, most of Russia's manorial and state industries relied on manual, traditionally organized labor. An average "factory" usually consisted of a number of artisan workshops. Most tasks were performed by skilled artisans—master foremen who were assisted by their apprentices. Many of these industries, especially the manorial ones, operated on a seasonal basis for only six to eight months a year, during the nongrowing season. According to Pazhitnov and other students of early Russian industries, the number of working days a year in these factories was about 250 to 260.[88] Some factories operated during the daytime and at night. An average workday in these factories lasted between eleven and twelve hours, workers employed in two five-and-a-half-hour shifts. Other factories operated only during the daytime, starting at five in the morning and continuing until eight in the evening, with a half-hour break for lunch. The workday in these enterprises was long and could last for thirteen hours or more.[89] Certainly, the seasonal character of these enterprises, as well as the workshop type of labor organization, determined children's labor conditions. It would probably be safe to assume that before industrialization, the majority of children employed in such businesses worked on an irregular basis and mostly performed ancillary tasks.

The extent of child labor and children's exploitation was perhaps higher in manorial factories than in state industries, especially when the market-oriented economy began to expand. For one thing, manorial factories for the most part remained free of state control and legal regulations. Thus, their workers, most of whom were not free, increasingly had to depend upon the will of the owner. As mentioned, some landlords who owned manorial factories sent whole families of indebted serfs to work in these enterprises and thus fulfill their dues and obligations.[90] According to some observers, because of the absence of state regulations, child labor in manorial factories sometimes took abusive forms. Legal records from manorial enterprises illustrate that in some instances, enterprises employed children for long hours and as regular workers—alongside apprenticeship or even instead of it. During the first half of the nineteenth century, most incidents of labor protest occurred in manorial factories. For example, in 1840 in the Wigel Textile Factory in Voronezh (central Russia), workers protested against children's employment. At this factory, children, especially juveniles, performed regular adult work and, together with adult workers, began work at three in the morning and worked until nine o'clock at night, with a four-hour break for lunch and rest. On average, children worked about fifteen to sixteen hours a day and received very low wages. Children's cheap labor in turn reduced the wages of adult workers, a development that caused the workers' protest.[91]

Legal records reveal another such incident. In 1842–43, about three hundred serfs, including many children, were ostensibly sent by their landlord to the Voskresensk Cotton Mill in the Dmitrov district of Moscow province in order to "learn the spinning industry." They in fact conducted ancillary work without any payment, as testified by the factory's workers when a strike occurred.[92] The employer and landlord insisted that "the children live [in the factory] in a quiet building, have healthy food, and perform effortless work suitable to their ages.... They have fresh faces, are laughing and healthy." The provincial officials found, however, that the children, who "still needed parental care," toiled at the mill day and night. Most worked in five-and-a-half-hour shifts that followed a six-hour break so that their total workday lasted about eleven hours.[93]

The investigation of this strike revealed that the serfs involved belonged to the nobleman Dubrovin of the Massal'sk district in Kaluga province, who had signed an agreement with Lepeshkin, the mill owner, and received forty rubles for each outfarmed person. In their testimony, the serfs stated that they were not gaining any training or education in the mill but were engaged in regular labor for which they received no wage. For example, Iakov Safronov testified that he had paid the landlord the entire 1844 rent of seventy rubles. The landlord, however, had sent him and his "underage" niece to the mill and promised that he would get four hundred paper rubles for every year that he worked in the mill. Lepeshkin, in fact, paid him nothing, stating that he had already paid the landlord for all workers. In the end, the landlord, in order to reach a compromise with the serfs, agreed to return some children to their homes and promised to compensate other serfs for their children with twenty-five paper rubles a year for each child. Although prohibited by the 1823 and 1825 decrees, the practice of farming out serfs and particularly serf children, according to contemporaries, was commonplace until 1861. Trying to evade the law, landlords indicated in legal documents that they had sent serfs and their children to factories as "apprentices to receive a professional education and training."[94]

Some episodes of child labor abuse occurred in state-owned industries as well. For example, in the state mines, eight-year-old children and elders over sixty—who, according to the law, were not supposed to work at all—sometimes engaged in "easy work," such as sorting and concentrating ore, carrying wood, and so on. According to the 1859 Orenburg provincial governor's report (northern Russia), the Pod'iachii Metallurgical Works used the labor of young children, elders, and persons with physical disabilities.[95] Although direct evidence of labor abuse and conflict in state factories and mines is scanty, many such episodes must have occurred.

In addition, it was not unusual for employers to assign apprenticed children to perform "ordinary" work done by adult workers.[96] Evidence from textile mills, for instance, illustrates that juveniles often worked as spinners and weavers. Their wages, however, were lower than those of adult workers, even when teenagers performed the same kind and volume of work.[97] In the Ekaterinoslavl' Stocking Mill, children were given the same work as adults but paid the lowest wage.[98] In the Altai region's mines and metallurgical works, children under age fifteen made copper cups, receiving only six rubles a year. For similar tasks, those between age fifteen and seventeen received twelve rubles annually, plus a daily bonus of two or three kopecks—much lower than the wages of adult workers. Of course, in general, children's productivity could not match that of adult workers, but their wages were significantly lower than their productivity. This was especially the case for teenage workers. Nevertheless, the production process in most industries involved numerous secondary and ancillary operations, and it was precisely these tasks that most children performed. In the Altai iron-ore mines, for instance, relatively few children engaged in cup production; most worked as auxiliary workers engaged in sorting and concentrating ore or other work "that suited their strength."[99]

Although historians debate the total number of hereditary serf workers employed in state and manorial factories, it is clear that from the beginning of the nineteenth century, prior to the 1861 reforms, this form of labor declined rapidly in favor of freely contracted laborers. This development affected many thousands of serf children. After the 1861 reforms, many such children moved from the countryside to become contract laborers in rapidly growing newly mechanized industries.

Early Tsarist Laws Regulating Children's Employment and Work

As noted earlier, during the early nineteenth century, state officials influential in the process of imperial lawmaking viewed child labor as a normal and useful practice aimed at apprenticeship and at preparing children for the responsibilities of adult life. As we have seen, child labor in state and some manorial industries was often regulated by specific statutes and decrees that applied to a single state or manorial factory. General mining statutes that governed labor in all mines and metallurgical works also existed. Meanwhile, most manorial and other private factories, namely those that had not been a subject of definite decrees, remained unregulated. For the most part, existing statutes dealt with bound or semibound serf labor. Moral acceptance of child labor was reflected in all these decrees, especially the practice of sending orphans, the urban poor, and hereditary serf children to state and manorial factories. The purpose of sending children to industries, as all these decrees stated, was to promote their education and welfare.

The earliest Russian decree that dealt in a general way with freely hired factory labor appeared in 1835. It was aimed at meeting the challenges of a rapidly expanding free market economy and securing a free labor force within the context of existing serfdom. The 1835 legislation demarcated the relationship between the employer and the employee, indicating that the employment of all workers in private industries rested upon the existence of a written personal contract between the two parties that clearly indicated the responsibilities of both sides. Although no provisions of this law concerned the employment of children directly, the law actually specified no age distinction and therefore applied to all persons, children as well as adults, who sought factory employment. Initially limited to Moscow and St. Petersburg and their districts, the decree had been extended by the early 1840s to most Russian industrial provinces.[100] During the 1830s, the government also introduced a series of decrees aimed at facilitating peasant mobility, which in turn helped bring a large number of rural children to factories.[101] Thus, the earliest decrees that dealt with free factory labor in effect legitimized the labor of children.

At the beginning of the nineteenth century, only a few humanitarian voices denounced child factory labor. During the 1810s, 1820s, and 1830s, the Russian government undertook some fragmentary measures to regulate children's employment, limited to certain industries and even single businesses. For example, in the late 1810s, the minister of the interior, O. P. Kozodavlev, proposed that the work of the wives and children of workers in state factories be outlawed.[102] He did so at the insistence of the local offices of the Interior Ministry, whose employees were often the first ones to deal with labor-related issues, hear workers' complaints, and record work-related accidents. The interior minister's concern for working children therefore in all likelihood reflected his awareness of poor labor conditions for women and child workers. Evidently, nothing came of this early proposal.

In 1835, in a message to Nicholas I, Finance Minister E. F. Kankrin suggested that employers be required to avoid employing juvenile workers for hard, laborious tasks and to limit their workday.[103] With the approval of the tsar, Kankrin issued a number of circular letters to industrialists' associations, commanding them to "provide welfare and education and not to exhaust [working children] with laborious tasks and for long hours, and [to] take into account the gender and age of each [child]." In 1835, the Moscow branch of the Manufacturing Council discussed Kankrin's suggestions and appointed a commission to inspect Moscow's factories.[104] Kankrin's efforts, however, met no positive response from the council and did not result in a substantial regulation of children's employment. The 1838 mining regulation reasserted that twelve-year-old children could be used only for auxiliary work and that only those over age eighteen could be employed as regular workers.[105] Again in 1843, the government issued a circular letter instructing employers to devote attention to conditions in workshops and to provide workers with living quarters and areas for rest. The letter required owners to assign working children easy tasks, appropriate to their physical strength and gender. It also suggested the need to provide schooling for working children.[106] During these decades, most state measures to improve labor conditions in industries were merely advisory, lacking any provisions for implementation. Consequently, for the most part, they did not produce significant results.

Obviously, these solitary acts and advisory recommendations were inadequate to restrict child labor decisively, especially in view of the fact that most state officials still viewed such labor as a form of education and apprenticeship.[107] In reality, most government officials did not see the labor of children in factories as a serious issue. In 1840, the British ambassador to Russia requested information from the Russian government about laws regarding child factory labor. The government replied that "since mechanized factories have not had substantial development in Russia, there are not many children working in the industries and there is no urgent need for labor regulation laws."[108]

The need for the introduction of a basic law regulating children's employment gradually became evident as government officials learned about widespread abuses of child apprenticeship in industry. Introduction of this legislation was actually provoked by the abovementioned workers' uprising at the merchant Lepeshkin's Voskresensk Cotton Mill in 1844. The events at the Voskresensk mill motivated the Moscow province's civil governor, Ivan Kapnist, to inspect the province's large factories. Concerned government officials thus found that children who were recruited as apprentices "to learn the spinning industry" were commonly assigned to regular labor tasks in most Russian industries and particularly in cotton-spinning factories.[109] During 1844–45, the Moscow government inspected twenty-three cotton mills and ten wool mills. According to its report, these factories employed about 2,100 children under age fifteen who worked day and night, twelve hours a day. In his report, the governor wrote that "although the machines make labor easier, night work cannot be easy for workers, and for children in particular, because of the character of the industry."[110]

Consequently, on August 7, 1845, the government restricted child labor in factories by prohibiting work between midnight and six A.M. for children under twelve years of age (see the appendix). The legislators placed the responsibility for implementing this law on local officials and factory owners and, unfortunately, did not introduce any penalty for its violation. In order to implement the law, the Moscow general-governor, A. G. Shcherbatov, required business owners to sign memoranda in which they promised to comply with the law's provisions.[111] It is not clear, however, that similar measures were taken in other provinces where industries employed many children. According to Mikhail I. Tugan-Baranovsky, an economist and student of Russian factories, employers in fact continued to evade the law, especially because legislators and local officials refused to establish an effective inspection system.[112] Additionally, this law made no provisions for providing juvenile workers with a school education. Still, regardless of its limitations, this was the very first meaningful attempt at legislation of child labor in Russia. It further signifies that long before Emancipation, and simultaneously with most industrialized nations, the Russian state had embarked upon the task of limiting the conditions under which children could be employed in industry.

The regulations of 1847 for state mines and metallurgical mills limited the workday for all children under fifteen in these industries throughout the country to eight hours. They required that these enterprises use the labor of children only in cases of exceptional necessity and assign them to easy work, according to their age and ability. Nevertheless, this law did not apply to private and manorial factories, where state control and regulations were lacking and where, according to some commentators, the workday for all workers, including children, could last for sixteen or more hours.[113]

These partial measures, limited to certain factories and industries, largely stemmed from the government's reaction to particular disturbances among manorial workers, which increased during the 1830s and 1840s. Consequently, these regulations of child labor were not uniform; they remained fragmented and specific. For the most part, the new laws focused on employment, age, and hours. They addressed no other forms of labor protection or working conditions in general. In essence, early tsarist decrees on child labor depended on the particular needs of concrete situations.

One problem arose from the fact that the early laws provided loose and quite flexible definitions of who was considered to be a child. For example, the law regulating the Urals' mining industry specified that "male children" under fifteen years of age were considered to be "underage" (*maloletki*), whereas fifteen- to eighteen-year-olds were teenagers (*podrostki*). At the same time, legislation for the Altai mines defined children under age twelve as "underage," whereas it defined those between twelve and eighteen as teenagers. These differences were important, since the definition affected children's actual employment. This legal flexibility resulted from concrete labor-force needs. To take only one example, the Altai mines, according to contemporary commentators, "had much work for children"—a factor that explains the region's less restrictive view of who was or was not underage.[114]

Some decrees established more precise definitions regarding the minimum age for employment, working hours, and workloads. For example, according to the statute regarding work arrangements in the Ekaterinoslavl' Mill, the workday for children was limited to twelve hours, in two six-hour shifts; thus, children worked twelve hours a day as adult workers. The statutes of the Tel'minsk State Wool Mill stated that the children of the mill's workers should begin work at the mill at age ten, laboring "according to their strength," and that fifteen-year-old children should accept regular full-time work.[115] The qualifier "according to their strength" signifies that their work was restricted to various auxiliary tasks. Thus, one may conclude that during the early nineteenth century, the standard age for beginning industrial employment in Russia was between ten and twelve, whereas the age of fifteen demarcated full-time employment; the standard workday was roughly twelve hours.

How do these early Russian laws on child labor compare with those of other industrializing nations of the period? Elsewhere in industrializing Europe, the first laws regulating children's employment were introduced in 1815 in Zurich, in 1819 and 1833 in Britain, in 1839 in Prussia, in 1841 in France, and in 1843 in the northern parts of Italy. In 1852, Sweden introduced similar legislation, while Austria did so in 1859. The laws set the minimum employment age (usually eight or nine, as in the British, French, and Prussian statutes); banned children of various ages from working at night; limited their daily work hours; and introduced factory inspectors to supervise the laws' implementation. The 1819 British act, which limited the employment age to nine and introduced factory inspectors, originally concerned only children employed in cotton mills. In 1833, the law was extended to the entire textile industry. By the mid-1850s, child labor laws had become common for most of industrializing

Europe. Like the Russian law of 1845, most of these early European laws lacked sufficient provisions for enforcement and were evaded by employers, as confirmed by historians of labor legislation.[116]

Within the broader comparative context of European nations, early labor laws dealt especially with child factory labor, while leaving aside other social groups of workers. With the exception of Britain—where the 1842 and 1844 laws prohibited underground work for children under age ten and restricted nighttime work for women and where the 1847 law limited the workday for women in the textile industry to ten hours—the employment of women was not yet a subject of specific concern. The early Russian decrees also did not address women's employment. Thus, the Russian law of 1845 and the earlier legislative measures place Russia within this general European tendency to protect working children only. The 1835 Russian statute, which introduced an employment contract and addressed other labor questions, reflects Russia's socioeconomic uniqueness, the decree's provisions mediating between an emerging free market and serfdom. Thus, one may argue that in Russia, as elsewhere, children were the category of workers who first came to the attention of the government as a group requiring state protection. The policy of state intervention in factory work and labor relations had its beginnings in attempts to regulate children's employment, a factor that further underscores the topic's significance.

Overall, in Russia, as in other parts of Europe, early factory legislation failed to establish uniformity in industrial labor legislation. For the most part, the laws remained fragmented and specific, addressing only certain industries or social categories of workers. Nor did these laws deal with workers' education and social welfare, except as regards certain poorly implemented statutes on child labor. Despite obvious shortcomings, the earliest European legislation, including Russian variants, had a positive side. First, it signified the readiness of politically diverse states to intervene in labor relations. Second, as will become clear in the following chapters, during the following decades, industrial labor in general became an issue of concern for state authorities and social reformers, beginning with the child labor protection acts. Both in Russia and elsewhere, the issue of workers' education and welfare dominated these debates. For all its limitations, the Russian law of 1845 signified the beginnings of a transformation of government officials' attitudes toward child industrial labor.

Doubtless, all these early attempts to regulate child labor had relatively little immediate effect on children's employment. In the middle of the nineteenth century, the perception of child labor as practical and morally acceptable still prevailed. The number of children working in industries continued to grow relentlessly, as did the number of new businesses that used child labor. In 1844, for example, there were about 3,000 children working in the industries of Moscow province, two-thirds in the cotton industry.[117] By the end of the 1850s, as peasant migration accelerated, the number of children employed in the industries of the province increased to 10,184, accounting for 15.2 percent of the province's industrial workers.[118]

Thus, well before rapid industrialization in Russia, child labor had been a widespread practice, a form of preparing children for adult life. It was welcomed by most social strata and supported by state laws. Because children's involvement in productive labor had long been a morally accepted custom, and sometimes because of the indispensability of children's wages to family income, parents were willing to send their offspring to emerging factories to gain an apprenticeship. Simultaneously, manufacturers viewed children as more adaptable to the new factory regime and more able to learn to work with new machinery and technology than adults. The conjuncture of these factors insured that children would remain an important source of labor for late nineteenth-century Russian industrialization.

2 ✦ Children in Industry
Demographic and Social Context

GREAT CHANGES occurred in the Russian economy during the middle decades of the nineteenth century. By 1850, a new capitalist mode of production had begun to challenge traditional manufacturing systems. Manorial and state factories showed signs of continued decline,[1] whereas free market enterprise began to expand rapidly.[2] The largest children's employer, the cotton industry, experienced the most remarkable development of all the industrial segments. The mechanization of the industry during the 1840s and 1850s—the earliest stage of Russia's industrialization—created a great need for unskilled and auxiliary workers.[3] The rapid development of new capitalist forms of production also spurred important changes in the employment system. In contrast to state and manorial factories, where hereditary serf labor dominated, and in contrast to domestic forms of manufacturing, which relied on the labor of family members, new capitalist enterprises employed contracted wage laborers. By the 1850s, free labor had supplanted bound labor as the prevailing type of industrial employment.[4] With these developments as a basis, during the second half of the nineteenth century, Russian industrialization entered a dynamic phase of expansion.

During this period, rapid population growth and changes in the rural economy complemented the accelerating tempo of the capitalist economy. After the 1861 reform, millions of rural residents, adults and children, sought industrial employment. Overall, the population of the empire increased from 73.6 million in 1861 to 131.7 million in 1900, although the most significant growth occurred in European Russia.[5] The urban population grew from 5.7 million in 1857 to 26.3 million in 1914.[6] This rapid increase in urban dwellers resulted mostly from peasant migration from the countryside. Facing economic hardship in the village, some peasant families moved to industrial centers where they hoped to find employment or better opportunities.[7] According to the demographic historian A. G. Rashin, the number of industrial workers grew from 706,000 in 1865 to 1,432,000 in 1890. Although frequently cited, such figures for the late nineteenth century cover only workers reported by factory inspectors and do not include large workforces in state metallurgical, mining, textile, and military industries; on railroads; in small factories and workshops; and so on. According to the 1897 census, industrial and agricultural wage workers accounted for 9,144,000 persons, including about 1,100,000 children under the age of fifteen.[8] Thus, children comprised a considerable part of factory and other forms of wage labor. During 1879–85, about 33 percent of the Moscow province's factory workers began their employment under the age of twelve, 31 percent between the ages of twelve and fourteen.[9]

What factors influenced children's factory employment during industrialization? What changes did industrialization bring to the traditional practices of child productive labor? Why did owners of newly mechanized industries employ children, and what kinds of work did children typically perform in these factories? In exploring these questions, it becomes clear that those whom scholars routinely generalize, sight unseen, as "the working class" consisted, along with adults, of large numbers of children, both boys and girls.

Statistics and Dynamics of Child Factory Labor during Russian Industrialization

In 1882, the medical doctor N. F. Mikhailov described one industrial giant: "When one approaches the factory building, this enormous 1,613-yard-long beast, one cannot even guess that the mouth of this animal absorbs a huge mass of children."[10] Mikhailov's admittedly melodramatic statement nevertheless brings to mind the striking reality that children represented a significant segment of the labor force. How many children actually entered factory labor during industrialization in Russia, and what were the characteristics of their employment? Estimating the number of children employed in industries during early Russian industrialization still presents a daunting task. Statistics on child labor are abundant but highly fragmentary and limited to certain industrial regions or to groups of individual factories. During the 1850s and 1860s, no statewide comprehensive survey of factory labor in Russia, much less child labor, had yet been conducted. The absence of systematic data and regular surveys of child labor reflects the government's lack of coherent concern about children's employment in industries at that time. As noted in chapter 1, the state did not view child factory labor as a serious social issue and continued to accept it as a means of teaching children industrial professions and preparing them for adult life—in other words, as apprenticeship.

The existing fragmented data from certain factories and some industrial areas suggest that by the mid-nineteenth century children age sixteen and younger comprised 12.0 to 15.0 percent of factory workers.[11] It is clear that with the expansion of the capitalist economy during the following decades, the absolute number, if not the percentage, of children working in industries rose dramatically. Available figures for industries in Moscow province, for instance, demonstrate that by the end of the 1850s, the number of child workers was 10,184, or 15.2 percent of the province's factory workforce.[12] In about ten years, in 1871, the number of working children had almost tripled, to 29,144, or 15.4 percent of Moscow province's 188,853 workers.[13]

In 1859, a St. Petersburg government commission chaired by A. F. Shtakel'berg studied labor conditions in St. Petersburg factories and gathered data on workers from 103 factories of the city and its district (*uezd*). These factories employed 16,224 workers, of which 1,282 (7.9 percent) were children age fourteen and younger. Most children, about 75 percent, were employed in textile factories, and 48.0 percent of these worked in cotton-spinning mills. The proportion of children employed in the cotton industry was probably even greater because many textile enterprises were recorded as "weaving" or "dyeing" mills, whereas some of them produced cotton goods as well. Children constituted 7.5 percent of cotton-spinning mill workers and 12.5 percent of laborers at weaving and dyeing enterprises. The highest proportion of children to adult workers was in print-type foundry mills (21.9 percent) and bronze works (18.1 percent), whereas children accounted for only 1.6 percent of the metallurgy industry's workers.[14]

Perhaps the fullest and most accurate data on St. Petersburg's factory workers for this period comes from the city's 1869 census. The census recorded 139,290 workers in the city's industries, among whom 13,587 (9.7 percent) were children age fifteen and younger. The breakdown of these statistics suggests that the capital's factories employed a relatively small, if still significant, number of very young children. For example, the city's industries employed 451 children age ten and younger and 747 eleven-year-olds (together making up 0.8 percent of the total workforce). At the same time, 4,636 children between the ages of twelve and thirteen (3.3 percent) and 7,752 between fourteen and fifteen (5.6 percent) labored in the city's plants. Juveniles (between sixteen and nineteen) were a larger component, comprising 14.0 percent (19,694) of the St. Petersburg labor force. Most of the children recorded in the census worked in cotton and tobacco mills. Although these statistics on early industrialization specify children's ages, they reveal few specific details about their gender and occupations.[15]

Detailed statistics on children's industrial employment first originated during the 1870s, when various state agencies and public associations began to gather data on child factory labor. In 1874, the Commission for Technical Education of the Russian Technical Society made an independent empirewide inquiry among industrialists and acquired considerable information regarding juvenile employment. Although many industrialists failed to respond to the inquiry,[16] the commission did receive data from 135 businesses, located mostly in the empire's industrializing provinces.

2.1 Data on Employed Children Gathered by the Commission for Technical Education from Some Central Russian Provinces in 1874

	Age Range					
	6–9	10–12	13–15	16–17	18	Total
Number	42	574	1,154	840	480	3,090
(%)	(1.4)	(19.0)	(37.0)	(27.0)	(15.6)	

Presented in table 2.1, the 1874 statistics generally confirm the data gathered in earlier surveys. In the responding businesses, 3,090 employees (17.8 percent of the workforce) were children and juveniles from six to eighteen years of age. The number of children employed in the surveyed enterprises ranged from 6.0 percent of the workforce in a rope factory to 40.0 percent in a hat factory. The youngest child worker was a six-year-old boy, and the youngest female workers were two eight-year-old girls. In these enterprises, girls accounted for 21.0 percent (649) of the 3,090 working children.[17] These data, although more detailed than in previous decades, are fragmented and must be used along with information from later surveys.

During the 1870s, local and provincial governments began to conduct surveys of factory labor on a regularized basis. This reflected increasing government and public concern about working children, a subject that will receive further discussion in the following chapter. A Moscow city government commission on factory labor organized in 1877 made inquiries about the city's industrial workers and found that in 1879, out of the 53,408 workers in Moscow industries, 2,077 (3.8 percent) were children under age twelve and 4,628 (8.7 percent) were juveniles between ages twelve and fifteen.[18] Most of these children (4,557, or 68.0 percent), worked in the textile industry, constituting 12.9 percent of the industry's labor force. Table 2.2 represents the data gathered by the Moscow government commission and shows the number of workers and children employed in Moscow industries in 1879.[19]

According to the data on St. Petersburg industries gathered in a similar survey in 1878 (presented in table 2.3), of 23,033 workers, 2,187 (9.5 percent) were children between ages ten and fifteen. Some observers noted that this data is incomplete and in fact represents only a small portion of the city's industries.

Nonetheless, it illustrates the general tendency and dynamics of child labor in St. Petersburg.[20] As in Moscow, the majority of these children (1,405, or 64.0 percent) worked in the city's textile mills. Although the proportion of children in metallurgy remained lower than that in textiles, their actual number and their percentage in comparison to most other industries were significant. Of interest is that by 1878 the percentage of children employed in St. Petersburg textile and metallurgical mills had increased markedly to 16.5 and 5.5 percent, respectively, since 1859, when children constituted, respectively, 8.8 and 1.6 percent of these industries' workers. By any measure, industrial growth during these decades was accompanied by a significant increase in children's employment. The statistics for St. Petersburg suggest that despite the proportional and absolute growth of child employment in textiles and metallurgy, the overall proportion of employed children to adult workers did not increase. This presumably signifies a drop in some other industrial sectors or, alternatively, mere statistical anomalies. As noted, data for Moscow and for all

2.2 Workers Employed in Moscow Industries in 1879

Industry	Number of Mills	Number of Workers	Number of Children					
			Number			Percentage		
			Under 12	12–15	Total	Under 12	12–15	Total
Textiles	306	35,348	1,723	2,833	4,557	4.9	8.6	12.9
Metallurgy and machine making	111	5,777	54	532	586	0.9	9.2	10.0
Food	69	5,569	167	504	671	3.0	9.0	12.0
Paper making and tanneries	38	2,373	46	245	291	2.0	10.3	12.3
Other industries	124	4,342	86	514	600	2.0	11.8	13.8
Total	648	53,408	2,077	4,628	6,705	3.8	8.7	12.5

2.3 Workers and Children Employed in St. Petersburg Industries in 1878

Industry	Number of Workers	Number of Children	Percentage
Metallurgy	9,018	502	5.5
Textiles	8,507	1,405	16.5
Ceramics	2,484	96	4.0
Food processing	1,067	26	2.6
Lumber	1,062	56	5.0
Chemicals	552	27	5.0
Paper	342	77	22.0
Total	23,033	2,187	9.5

of Russia suggest increases in the absolute numbers of child laborers, with no significant proportional increase.

The widespread use of child labor during the 1870s is also illustrated by data from individual businesses. For example, in 1878, the Morozov Textile Mill in Tver' province (central Russia) employed 4,536 workers, including 736 children under age fifteen and 1,198 juveniles between fifteen and eighteen (16.2 and 26.4 percent, respectively). In the same year, 720 (20.0 percent) of the 3,600 workers of the Rozhdestvensk Textile Mill (Tver' province) were children and juveniles.[21] In the late 1870s, the Iartsev Textile Mill in Smolensk province (southwest Russia) had a workforce in which children between ages seven and fourteen constituted about 25.0 percent of the total. At the same time, juveniles between fifteen and eighteen accounted for 26.0 percent of the mill's workers, adults only 49.0 percent.[22] Although these proportions are significantly higher than those shown in general statistics, the possibility cannot be excluded that, because of employers' natural tendency toward concealment, they represent the proportion of children employed in the textile industry more accurately.

The first coherent nationwide census of child labor in Russia occurred in 1882. The Ministry of Finance's Department of Commerce made inquiries through its local agencies and its newly created factory inspectorate (1881) about the employment of children in private industry (see chapter 4). By August 1883, 2,792 manufacturers across the empire had responded. Although the 1882 data does not by any means represent all private factories, this was nevertheless by far the most comprehensive survey of Russian private factory labor to date. Table 2.4 displays the number of workers and children employed in the 2,895 reporting factories in 1883.[23]

Although it is hardly possible to know the exact number of children employed in all Russian industries in 1883 or any other year, the figures presented in table 2.4 shed light on important aspects of child factory labor. The figures show that 49,581 (9.2 percent) of the 540,794 factory workers reported were fifteen years of age and younger. The overwhelming majority of these child laborers (30,171, or 60.9 percent) engaged in textile production, in particular in the cotton industry (18,826 children, or 38.0 percent)—a tendency also suggested by earlier surveys. Cotton- and wool-spinning mills employed very high percentages of children, 21.0 and 31.2 percent respectively. Private mines and metallurgical works also employed large numbers of children (7,667, or 15.5 percent). Many children also worked in food-processing mills (6,458, or 13.1 percent). The industries defined as "other" included those that processed organic materials, such as animal skins and bones.

The data on child labor gathered in 1883 contain separate figures for boys and girls employed in the reporting factories. The number of girls age fifteen and younger is significantly lower than the number of boys of the same age. In Moscow's factories, for instance, out of 1,756 children age fifteen and younger, 1,451 (82.6 percent) were boys and only 314 (17.4 percent) girls. For children age sixteen and older, the percentage gap between boys and girls decreases somewhat. For example, of 1,320 juvenile workers between sixteen and eighteen, 985 (75.0 percent) were males and 335 (25.0 percent) females, whereas men and women from age nineteen to fifty accounted, respectively, for 6,214 (78.0 percent) and 1,746 (22.0 percent) of the workforce.[24] In Vladimir province, men comprised 63.7 and women 36.3 percent of the workers.[25] This tendency is also reflected in the data from earlier decades. Most girls remained in the coun-

2.4 Number and Percentage of Workers and Children Employed in Various Reported Industries in 1883

Industry	Number of Mills	Number of Workers	Number (%) of Children			Total Number (%) of Children
			Under Age 10	10–12	12–15	
Fiber processing						
Cotton spinning	27	14,935	56 (0.4)	406 (2.7)	2,666 (17.9)	3,128 (21.0)
Cotton weaving	40	22,929	99 (0.4)	534 (2.3)	2,087 (9.1)	2,720 (11.8)
Cotton finishing	76	36,279	25 (0.1)	446 (1.2)	3,423 (9.4)	3,894 (10.7)
Other cotton-processing mills	31	80,779	68 (0.1)	1,371 (1.7)	7,645 (9.5)	9,084 (11.2)
Linen spinning and weaving	18	22,251	46 (0.2)	738 (3.3)	2,948 (13.6)	3,732 (16.8)
Other linen-processing mills	20	1,987	0 (0.0)	2 (0.1)	76 (3.8)	78 (3.9)
Wool washing	16	4,872	128 (2.6)	207 (4.3)	570 (11.7)	905 (18.5)
Wool spinning	22	3,568	3 (0.1)	115 (3.2)	995 (27.9)	1,113 (31.2)
Wool weaving	32	10,092	14 (0.1)	142 (1.4)	659 (6.6)	815 (8.1)
Wool cloth making	103	25,135	44 (0.2)	537 (2.1)	2,417 (9.6)	2,998 (11.9)
Other wool-processing mills	10	899	0 (0.0)	0 (0.0)	22 (2.4)	22 (2.4)
Silk weaving	18	4,288	7 (0.2)	53 (1.2)	288 (6.7)	348 (8.1)
Other fiber-processing mills	89	9,719	24 (0.2)	167 (1.7)	1,143 (11.8)	1,334 (13.7)
Total fiber-processing mills	502	237,733	514 (0.2)	4,718 (2.0)	24,939 (10.5)	30,171 (12.7)
Mining and metal	709	145,053	55 (0.6)	404 (0.3)	7,208 (5.0)	7,667 (5.3)
Food processing	811	105,726	154 (0.2)	848 (0.8)	5,456 (5.16)	6,458 (6.1)
Minerals	209	15,003	142 (1.0)	688 (4.6)	1,767 (11.8)	2,597 (17.3)
Lumber	240	17,649	46 (0.3)	114 (0.6)	933 (5.3)	1,093 (7.9)
Printing, binding	79	3,536	0 (0.0)	17 (0.5)	609 (17.2)	626 (17.7)
Chemicals	142	8,172	0 (0.0)	86 (1.1)	505 (6.7)	591 (7.8)
Other industries	203	7,922	50 (0.6)	66 (0.8)	262 (3.3)	378 (4.8)
Totals	2,895	540,794	961 (0.2)	6,941 (1.3)	41,679 (7.7)	49,581 (9.2)

tryside, while some left to work in various domestic services.[26] As noted in the previous chapter, many of those who remained in the village engaged in cottage industries in addition to their numerous household and agricultural activities. For all its personal, social, and economic significance, none of this labor appears in statistics.

Regardless, the statistics of table 2.4 reveal children's employment in large and medium-size private factories with sizable workforces—that is, those factories that reported to the commissions and were visited by factory inspectors. These statistics neglect entirely state-owned businesses and, perhaps even more important, small private enterprises and services that also employed numerous children. In addition, factory inspectors had no access to certain distant enterprises, which, therefore, are not covered by these statistics. According to one of the chief factory inspectors, Ia. T. Mikhailovskii, factory inspectors reported on only about 20 percent of all private businesses. These were, however, large and medium-sized mechanized enterprises, which presumably employed the majority of workers.[27] According to V. I. Lenin's estimates, in the early 1880s, businesses that employed more than 100 workers accounted for about 5 percent of all businesses in Russia but utilized the labor of fully 67 percent of wage workers overall. In 1890, large mechanized enterprises made up about 8 percent of the total and employed 71 percent of factory workers.[28] Thus, presumably, the majority of working children toiled in large and medium-sized mechanized enterprises, a reality that does not reveal the numbers of children employed in two numerous state-owned industries.

Historians often cite the figure of about 2,000,000 to indicate the total number of factory workers in late nineteenth-century Russia. This, however, takes into account only the data collected by factory inspectors and hardly accurately represents the actual number of industrial wage laborers. As noted above, according to the 1897 census, there were 9,144,000 wage workers in Russia, including agricultural wage earners. This figure includes 238,000 children (2.6 percent) below age twelve and 363,000 (4.0 percent) between the ages of thirteen and fourteen. Accordingly, children age fourteen and younger comprised 6.6 percent of wage workers. Juveniles between age fifteen and sixteen years totaled 644,000 (7.0 percent), those between seventeen and eighteen 1,181,000 (12.9 percent).[29] Thus, the proportion of children rose significantly with each incremental increase in age. Statistical data also suggests that most if not all industries used child labor.

It is likely that, in general, children age fifteen and younger constituted some 9–12 percent of Russia's industrial labor. Depending on industry and individual enterprise, however, the percentage may have ranged considerably. With the growth of the industrial economy during the late nineteenth century and the increased demand for wage workers, the actual number of children employed in industries grew rapidly. As mentioned previously, even in 1897, some fifteen years after the enactment of the first important child labor law in 1883, about 1.2 million children age sixteen and younger (19.5 percent of all workers) worked in industrial concerns. The 1882 law decisively restricted labor and apprenticeship for children under age twelve and regulated the employment of those between twelve and sixteen (see chapter 4).

Causes of Child Factory Employment during Industrialization

Why did so many children enter the factory labor force during industrialization? Why did industries employ children? Even before industrialization, as we have seen, society had broadly accepted child productive labor as a means of preparing children for adult life. Children usually engaged in productive labor in order to receive an apprenticeship and gain a professional education. Of course, such factors were also crucial in influencing children's factory employment during industrialization. But industrialization itself produced new economic and social realities that spurred child industrial labor. Thus, answers to these questions probably lie, on the one hand, in the dramatic economic and social changes in the countryside during the second half of the nineteenth century, which included the acceleration of the market economy, the rapid growth of rural population, and the decline of the traditional extended family. On the other hand, rapidly growing industries created a massive demand for wage labor. Further, child productive labor enjoyed wide popular acceptance, and there were no effective child labor regulations in private factories. All these factors created a situation in which children continued to be an easily available and even desirable source of labor for late nineteenth-century industrialization.

For the most part, Soviet historians followed certain nineteenth-century observers in emphasizing economic motives as the primary basis for child labor. This interpretation implied that capitalist enterprises mercilessly exploited low-paid child laborers in order to gain as much profit as possible. This view perhaps received its strongest reflection in the words of historian Vladimir Gessen, who stated in 1927 that "the cheapness of child labor [was] the stimulus for its broadest exploitation."[30] Of course, economic exploitation may have been a crucial factor, but it was by no means the only cause of a phenomenon that brought hundreds of thousands of children to enter factory employment.

Most contemporary observers of factory labor instead suggested multiple economic, technological, and social issues as the basis for widespread children's employment. In their view, owners of factories preferred child labor for several interconnected reasons, including economic ones.[31] Some commentators noted that mechanized factories favored the employment of children because, unlike in traditional crafts, work on the new machines often did not require specific skills or great physical strength. Mechanized production involved tasks that, in the eyes of many entrepreneurs, could be performed without special training or skill. If possible, they preferred to use child labor for such tasks.[32] In addition, contemporaries remarked that manufacturers viewed children as more adaptable than adults to the new factory environment, better able to learn to work with the new machinery and technology, and often better fitted physically to perform certain operations. Some industrialists even claimed that without children's input, adult workers could not accomplish many aspects of production at all.[33] These assertions find support in many contemporary accounts. One observer, who studied children employed as doffers in a spinning factory, wrote that "this task was not difficult. When bobbins are full, the doffer replaces them. The bobbins are located low, suitable only for children's height. Adult workers could hardly accomplish this task." In this view, the new industrial technology and the growing number of mechanized factories created a huge demand for child laborers.[34]

Other contemporary commentators insisted on economic factors as the chief motivation for the increased use of child labor, a factor also emphasized by Soviet scholarship. "The mechanized factory is extremely interested [in child workers]," wrote A. Romanov, a physician who investigated factory labor in 1875, "because it pays children less than adult workers. At the same time, [overall] wages can be lowered. Even adult workers have to accept these rates in the face of strong competition from young workers.... The factory administration forces out adult workers and replaces them with children. By keeping labor rates low, the owner tries to gain as much profit as possible."[35] Many factory inspectors shared this view. For example, Dr. P. A. Peskov, in his reports on factory labor in the Vladimir industrial district, maintained that the use of child labor in industries was caused "mainly by economic reasons—by the cheap labor of children." Peskov stated that there was no other reason to use child labor in factories, "because most of the tasks children performed could easily be accomplished by adult workers."[36] Peskov's last remark, it should be noted, contrasts sharply with the opinions noted above that emphasize aspects of the new industrial machines—that is, technological factors—as an important cause of accelerated child factory employment.

Soviet scholars of child labor suggested that the economic motive of hiring cheaper labor was particularly important for small, traditionally organized workshops, which relied mostly on manual labor. According to Gessen, "given their low technological level, artisan workshops could compete successfully with the mechanized factory only by employing children and keeping wage rates down."[37] In support of this assertion, Gessen pointed out that St. Petersburg industries employed fewer children than industries in Moscow because, allegedly, the former were more technologically advanced and did not have a great need to hire children.[38]

At first glance, this observation appears valid. For instance, the data from 1878 and 1879 for Moscow and St. Petersburg cited in tables 2.2 and 2.3 seems to support Gessen's conclusion. The data shows that in St. Petersburg, children accounted for 9.5 percent of the city's industrial workers, whereas in Moscow they made up 12.5 percent. But closer analysis raises doubts. The figures reveal, for instance, that textile factories in St. Petersburg had workforces that consisted of 16.0 percent children, whereas in Moscow textile mills children accounted for only 12.0 percent of workers. Yet the textile mills of Russia's "northern capital" were the most technologically advanced in the country, surpassing textile mills in Moscow and the other central provinces.[39] A further complication arises from the distinct possibility that differences between Moscow and St. Petersburg may simply reflect the data's incompleteness. Likewise, the 1883 data also indicates that many children worked in the cotton industry, the most mechanized of all textile industries. Children, for instance, comprised 21.0 percent of workers in cotton-spinning mills (see table 2.4). In view of these anomalies, the safest conclusion is simply that child labor, to one degree or another, was economically desirable both for new mechanized factories and for traditional manufacturing workshops.

Many contemporary observers also stressed that child labor resulted from the economic and social strategies of peasant families, an approach that has the merit of establishing peasant/working-class agency. These commentators noted that, driven by economic and social goals, parents were frequently willing to put their offspring to work in factories. The Finance Ministry commission maintained in 1860 that the "custom of putting children to work" was widespread.[40] According to Romanov, "the enormous proportion of children employed in the factory... suggests the strong need of the population; [parents] ... have decided to take the opportunity to turn into profit the physical feebleness of their children.... Children's employment provides these families with a valuable financial contribution."[41] In 1878, the Vladimir provincial governor noted that some parents were "tempted by the small profits" they could gain by sending their children to factories.[42] Likewise, some manufacturers emphasized parents' requests to explain why they employed children. One factory owner claimed that he "employed children only as a favor and deigned to concede to their mothers' humble requests.... As for me," he continued, "I do not need them at all."⁴³ Whether this industrialist's assertion of his humanitarian motives was sincere or not, his remark about parents' desires seems credible. Entrepreneurial motives and demeaning comments about parents' supposed greed aside, children's factory employment usually reflected the desires, needs, aspirations, and decisions of families and parents.

Nevertheless, as before industrialization, some parents continued to send their children to factories as apprentices with the purpose of gaining them a professional education. In such cases, parents concluded written or, more usually, oral agreements with employers that specified apprenticeship conditions. Apprenticeship usually lasted from two to five years, and in most cases, apprenticed children did not receive wages. Children gained training in various industrial professions—engraving, drawing in pencil and on pantograph, carving on wood in printing factories—or gained skills in working on the lathe and other machines, and so on.⁴⁴ For their parents, the major goal of their children's employment was preparation for adult life. Their children's apprenticeship usually made no immediate financial contribution to the family. Nonetheless, during the decades that followed the 1861 serf emancipation, as the economic needs of working families deepened, the traditional cultural acceptance of unpaid child labor in apprenticeships helped pave the way for paid child labor. Industrial apprenticeship no longer seemed to serve as a crucial factor for most children's entry into productive labor. Many parents put their children to work in factories primarily because of the family's economic need, rather than to train children for an industrial profession, although the latter also certainly occurred.

Both local government officials and late imperial scholars explored the economic conditions of lower-social-strata families and estimated how these economic conditions influenced children's employment. The St. Petersburg government, for instance, gathered data on a number of the city's working families during the early 1880s. In most cases, the data suggests that children's factory employment seemed to be determined by their families' economic needs.⁴⁵ In his seminal book on factory labor, the historian M. Balabanov cited the following example of a peasant family that had migrated to St. Petersburg in the late nineteenth century. The family consisted of six members, including a mother, a father, and four children. The mother did not work because she took care of the youngest baby daughter. The father found employment in a calico factory, where he earned twenty rubles a month. Balabanov noted that since the family could not sustain itself for a month on twenty rubles in St. Petersburg—a city where living was costly—the parents sent their older daughter to the factory. She earned seven to eight rubles a month, a significant contribution to the family budget.⁴⁶ E. N. Andreev of the Russian Technical Society described a similar case of a peasant family that had moved to St. Petersburg in the early 1880s. This family included four members: a father, a mother, and one male and one female child. The father and his twelve-year-old son worked in a factory where they together made thirty-eight rubles a month, whereas the mother and daughter stayed home. The son's share of the wages was about eight rubles a month. The boy worked twelve hours a day and could no longer continue the schooling that he had received before moving to the city.⁴⁷

Doubtless, not all families who put their children to factory work were needy. In some recorded cases, children from relatively well-off working families found employment in factories as well. For example, a St. Petersburg government inspector described a family of three—a father, a mother, and an eleven-year-old daughter. All three worked in a St. Petersburg mill and made a total of forty-six rubles a month. The daughter's monthly wage was about six rubles. This family spent about thirty-five rubles a month on its subsistence, accumulating about ten or eleven rubles as savings. The family had a land allotment in the countryside but granted the right to use it to one of their relatives, who in return was obliged to pay all local and communal dues and taxes for the family. The government inspector pointed out that this family had no acute economic necessity for sending their daughter to a factory but did so sheerly out of a desire to increase its income.⁴⁸ Thus, for some families, child labor arose out of aspirations for a better life in the present or future.

Regardless of the "temptation for a little profit," as one commentator put it, and the desire for a better future life noted by some contemporaries, economic need and poverty were the driving forces behind most parents' decisions to send their children to factories. Contemporary scholars estimated that, in order to subsist in St. Petersburg, a family of two adults with two or three children needed an income of at least twenty-five to thirty rubles a month. Even the simplest foods were expensive. The monthly cost of basic food necessities in a large Russian city during the late decades of the nineteenth century was four to six rubles per adult person.[49] In addition, an individual family made outlays on housing and other basic items, such as soap, kerosene, candles, and so on. The cheapest living space in St. Petersburg at the time cost about one silver ruble a month, whereas an average space cost three or four rubles. Consequently, factory employment was crucial for older children of most families with dependent children, such as those cited above. One factory inspector recorded in his report that a family with "many dependent mouths who could not work was happy to gain every little kopeck" earned by an older child.[50]

Scholarly studies have shown that nuclear families of two adults with children were not rare in rural Russia, and after the 1861–64 reforms, as the market economy accelerated in the countryside, their number grew considerably. Naturally, and concomitantly, the traditional extended family began to decay. Anthropologists and historians of the family have noted the general decline of the traditional multigenerational household and the growth of nuclear families in Russia (as elsewhere). According to the historian I. I. Milogolova, "in these new conditions [created by capitalism], the small family that consisted of parents and children began to prevail."[51]

Some late nineteenth-century observers suggested that, unlike in the traditional multigenerational household, "in a small nuclear family, persons can realize their essential desires for independent living and for working exclusively for the well-being of their own families."[52] This desire to live independently and well increased the economic pressures on individual members of such families. As noted in chapter 1, even before industrialization, the labor pressure on children in nuclear families usually exceeded that in extended families. During industrialization, factory employment of children often made possible the survival of nuclear families, when, repelled by harsh economic conditions on the land, they moved from the countryside to industrial areas. The fall of the extended household and the rise of the nuclear one seem to have affected the increase of children's factory employment. In addition, the rapid growth of the rural population of the Russian Empire during the second half of the nineteenth century affected the economic conditions of many peasant families. In 1858, the rural population of imperial Russia (without the Kingdom of Poland and the Duchy of Finland) was 68 million; by 1897, it had increased to 116 million. In 1913, the population of rural Russia reached 163 million.[53] According to the demographic historian V. M. Kabuzan, in European Russia, this growth occurred because of a decline in mortality rates and a simultaneous rapid increase in birth rates. Some demographers suggest general overpopulation in the countryside, along with an increased proportion of children in the general population.[54] In this interpretation, this resulted in a sharp rise of the number of families with small dependent children. Historians of child labor in industrializing Europe and America have noted that the presence of small children in a nuclear family increased the pressure on older children to engage in wage labor.[55]

The rapid rural population growth spurred temporary peasant migration to urban centers, which were growing significantly by the late nineteenth century. Agrarian historians find that about 12.5 million peasants, including many children, annually migrated temporarily to industrial areas during 1900–1910.[56] This figure, however, is based on the number of documents given to peasants for temporary leave and therefore may not fully represent the actual number of peasant migrants. In many cases, peasants migrated without these documents. In order to find employment outside the village, peasants moved individually or in work units known in Russia as *arteli*. These units usually included from four to twelve people, both male and female adults and children. Sometimes, however, they consisted of only one gender or only of children. Rural children most commonly migrated with their families, fathers, or other adults. In some cases, however, they joined *arteli* led by older children.[57]

In fact, many peasants did not have to leave their villages in order to get factory employment. According to economic scholars, many industries in Russia were located in the countryside. In Vladimir province, for instance, all factories were situated in the local districts (*uezdy*) or near villages rather than in cities. The city of Vladimir had no factories and only a few artisan workshops. The nearest factory, the Nikitin Cotton Mill, was located in the Lemeshki village, about sixteen kilometers from the city.[58] This tendency probably finds its best expression in V. I. Lenin's words: "If the peasant does not go to the factory, the factory does go to the peasant."[59] Thus, the rapid growth of the rural population created economic and social conditions that increased the pressure on children to seek factory employment.

Furthermore, all the factors that influenced the entrance of children into wage labor during industrialization were complemented by the absence of effective labor regulation laws. As we have seen in chapter 1, children's employment in private businesses remained largely unrestricted by law. In practice, the employment of children was as easy as the employment of adults. During industrialization, children continued to be employed in factories much as they had long before the 1861 reforms. Employment contracts were often informal and oral. In many cases, children were hired not by the business administration but by factory foremen who worked under the administration. Thus, one may suggest that the participation of children in the labor force during industrialization resulted from a multiplicity of economic, social, and cultural factors. These included the broad popular acceptance of child labor, the economic and technological interests of entrepreneurs, the growth of the child population, the economic pressures on families, and the absence of labor protection laws. All these factors worked together to make children an important part of the labor force during industrialization.

Employment, Work, and Living Conditions of Factory Children

How did industrialization affect children's employment conditions? What tasks did children perform in factories? Children's employment in industries and the type of labor they performed seem to have been determined by many factors, including the size and character of the business, its location, and various local and individual enterprise arrangements. By no means did industrialization bring about an immediate break with all the old customs of employment and work. To the contrary, there was a considerable degree of continuity. Many traditional practices of children's employment and family labor continued. During the late nineteenth century, some industries in Russia still primarily used manual labor. They were traditionally organized and operated on a seasonal basis, for only six or eight months a year. Although the number of such enterprises was declining sharply, to the advantage of new mechanized ones, the proportion of children compared to the total number of workers in these traditional enterprises was quite high, sometimes reaching 40 to 50 percent. Even so, as noted, most employed children worked in mechanized factories.

Bast-matting workshops, for instance, provide an excellent example of employment practices and work organization in small businesses in Russia. These shops relied on family labor and operated only six to eight months a year, during the nongrowing season. Workers were organized in teams called *stany*. Each team (*stan*) consisted of four people, usually members of one family, and worked on one bast-matting frame. An adult male, usually the father, who was called "the standing person" (*stoiachii*), operated the machine. He was helped by an assistant (*zarogozhnik*), an adult man or more likely a boy of age fifteen or sixteen. They were further assisted by a helping boy (*zavodiashka*), a child between ten and fifteen years of age, who prepared the bast warp and performed other tasks. The fourth team member, an adult woman (*chernovakha*), usually the mother, carded the bast. She also prepared food for the team. Thus, children and juveniles might constitute about half the workers in these bast-matting mills.[60] Although others probably existed, evidence indicates that nine bast-matting factories were located in Moscow in 1882 (including five in Moscow province); these concerns employed over 2,000 workers, about 33 percent of whom (660) were children age fourteen and younger.[61] Some descriptions specify that these enterprises also used the labor of very young children of six, five, and even three years of age.[62]

Labor organization in bast-matting workshops remained much as it had been in the early nineteenth century. In 1897, the medical doctor E. M. Dement'ev described the work organization in one bast-matting workshop:

The whole team starts its work at 4 o'clock in the morning and makes its first round of 7 matts by 8 A.M. After that the team workers have their breakfast while still continuing to do some work. After 8 o'clock the *stoiachii* [standing person] takes a rest and the assistant takes his place, while the helping boy [*zavodiashka*] stands in for the assistant. Having slept through the making of 5 matts (2.5 or 3 hours), the *stoiachii* again sets to work while the helper boy takes a rest also for 5 matts. By 2 P.M. they finish the second round of 10 matts and then all have a 30-minute lunch. Then the *stoiachii* and the helping boy take a rest for 2.5 to 3 hours. From 8 P.M. all four work together and by 2 o'clock in the morning make 10 matts more. Then the team has its dinner and rests.[63]

Thus, the team in bast-matting mills worked the whole day, with three two-and-a-half- or three-hour breaks for each member of the team. Presumably, the woman assistant (*chernovakha*) and the helping boy probably had more time to rest.

The employment process in these factories was carried out much as it had been during preindustrial times. Dement'ev observed that owners of bast-matting factories in Moscow province "annually, at the end of summer or the beginning of autumn, send their personnel agents to the Moscow district where they in turn recruit their workers for next year through their trusted local agents [*riadchik*].... Employment agreements are concluded not with the individual worker but with the *stoiachii*, who would then need to locate and hire his own assistants." The *stoiachii* usually employed members of his family.[64]

This kind of family employment arrangement was certainly not unique to bast-matting workshops. Evidence suggests that many children were in fact hired not by manufacturers or the factory administration, but by individual workers whom they assisted. According to factory inspectors' reports, factory administrations sometimes did not even know how many children they employed because children were recruited by foremen as individual workers, who hired their own children or the children of their relatives. For example, one Vladimir province cotton factory employed twenty-nine dye grinders—children between eight and fifteen years of age—who assisted adult male hand dyers. Of the twenty-nine children, twelve worked for their own fathers or other relatives (uncles or older brothers), and the remaining seventeen worked for nonrelated adults.[65] This practice also occurred in mechanized industries, but to a lesser degree. In spinning mills, for instance, some children helped their fathers who worked as spinners. According to Dement'ev, during 1874–84, about 50 percent of factory workers were sons and daughters of persons who worked in factories. However, he specified neither the ages of these workers nor the type of their work.[66]

Regardless, a majority of employed children evidently assisted nonrelated adult workers. Thus, although a preindustrial tradition of family employment still persisted in some enterprises during the late nineteenth and early twentieth centuries, it was in decline. It should be mentioned that under this type of labor organization, children usually worked under the supervision of their parents or other family members. Especially for small artisan-type enterprises, industrialization brought only slow change to family labor as traditionally practiced in the countryside. Child labor there remained much as it had been for centuries.

The figures cited in the previous section demonstrate that many children worked in new mechanized industries, where the majority of wage workers overall worked. The cotton industry, for instance, was far and away the most mechanized of the various textile industries and employed the largest number of child laborers. The 1874 Commission for Technical Education figures illustrate that 22.4 percent of the cotton-industry labor force consisted of child and juvenile laborers from six to eighteen years of age. The figures for 1883, presented in table 2.4, suggest that about 61.0 percent of children employed in industries and re-

2.5 Ages of Child Workers in Workshops of the Sokolovskaia Cotton Mill (1882)

Mill's Department	Number of Child Workers									Total			Number of All Workers		
	Under 10			10–12			12–15								
	Male	Female	Total	Male	Female	Total	Male	Female	Total	Male	Female	Total	Male	Female	Total
Spinning	1	2	3	6	13	19	90	12	102	97	27	124	342	236	578
Weaving	0	0	0	1	2	3	19	81	100	20	83	103	318	572	890
Printing	0	0	0	3	0	3	46	0	46	49	0	49	976	301	1,277
Total	1	2	3	10	15	25	155	93	248	166	110	276	1,636	1,109	2,545

ported by factory inspectors worked in textile factories. Children fifteen years old and younger accounted for 12.2 percent (18,826) of the industry's workforce. As mentioned earlier, mechanized production included many operations that did not require special skills or strength and could be performed by persons with little or no training. With the expansion of mechanized factories, the demand for such workers increased further. As noted by many contemporaries, most children who worked in factories engaged in ancillary activities, including running errands.

The pervasiveness of child labor in textiles was by no mean confined to Russia; it was typical for all industrializing countries. For example, evidence from the British Parliamentary Papers for England show that in 1874, of the textile workforce, 12.5 percent were children between eight and twelve years of age, 8.4 percent were male juveniles between thirteen and seventeen, 54.4 percent were women thirteen and over, and 24.7 percent were men eighteen and over. Overall, the British cotton industry employed a larger number of children than did other industries.[67] Likewise, in 1865, most of France's child laborers (59.7 percent) were employed in textile (mostly cotton) mills.[68] A very similar pattern existed in the southern portion of the United States during the late nineteenth century: more than 60.0 percent of working children in the South were employed in the region's cotton mills.[69]

What work did children typically perform in cotton mills? The profiles of child labor in the industry are suggested by table 2.5, which presents data about child workers in the main workshops of the Sokolovskaia

Cotton Mill in 1882.[70] The Sokolovskaia Cotton Mill, owned by A. Baranov, was located in the Aleksandrovskii district (*uezd*) of Vladimir province, the center of Russia's textile production. The mill had several main workshops, which included spinning, weaving, and finishing shops, as well as secondary works—an iron foundry, a metal workshop, a brickyard, and a peatery. The total mill workforce consisted of 3,496 workers—2,221 male (63.5 percent) and 1,275 female (36.5 percent), all local peasants. Of the 2,545 main workshop employees, 276 (10.8 percent) were children under fifteen years of age.

Many children worked in the cotton industry because the cotton production process included many operations and tasks that required unskilled or semiskilled labor, as well as ancillary activities. Table 2.5 illustrates that 45 percent of the Sokolvskaia mill children (124) worked in the spinning shop. Most of these children performed auxiliary operations including setting up bobbins (62 children), sorting (13), and other secondary tasks such as cleaning machines and floors, and so on. Some children were assigned to piece together broken threads (19 children), which ordinarily would be a task assigned to spinning-machine operators. The new mechanized process of spinning associated with the introduction of the self-acting mule created a demand for semiskilled and unskilled workers to assist spinners. All 32 spinners in the shop were male adult workers because the operation of the self-actor required strength. In the weaving shop, 51 child workers between ages twelve and fifteen were weavers, 52 mostly younger children—secondary workers. The printing and dyeing department of the Sokolovskaia mill employed 49 children. Most children who worked in these departments also performed auxiliary tasks, such as grinding dye, cleaning equipment, and carrying things.[71]

These figures on children's occupations in the Sokolovskaia mill conform to general tendencies in the cotton industry. Contemporaries noted that men in cotton mills usually performed jobs that required greater strength, whereas women and children performed "easier" tasks, which nevertheless were often dirtier and more dangerous. According to factory inspectors, many children employed in cotton-spinning mills worked in preparatory facilities on carding and scutching machines.[72] In weaving rooms, children usually worked as helpers. Their job consisted of putting warp through the openings of reeds. The helping boy (*podaval'shchik*) passed the thread through reeds, and the boy who assisted him took hold of it, completing the process. Many children also worked on spooling and winding machines. Children who worked as weavers usually had their own looms and performed all the tasks of adult workers.[73] In the dyeing and printing departments, most children worked on drying and starching drums. There they usually ensured that the cloth did not jam when it went onto the drum.[74] Many children served as the assistants of workers who operated dyeing and washing equipment. They also carried calico and engaged in the final tasks of cloth finishing. In all these shops, children also cleaned machines and equipment, carried products, wiped floors, and performed other errands.

Thus, children younger than twelve employed in textile mills normally assisted adult workers (sometimes their own fathers or other relatives) and performed ancillary tasks. A few children, usually between the ages of twelve and fifteen, performed adult jobs. While helping adult workers, children learned to work with machines on which they could replace adult workers in a few years. One contemporary noted that children age ten to twelve "observed the work of adult workers and tried to imitate it.... The most active children helped spinners, and they learned how to piece together broken pieces of yarn and so on." When they had learned all the operations required for spinners, they began to work as spinners themselves.[75]

How did industrialization affect working hours for children? Before the introduction of labor regulation laws, the workday in fact changed very little and in most cases remained as it had been before industrialization. Working hours for children usually depended on labor arrangements in individual factories and were usually the same as for adult workers. Most, if not all, textile mills that employed large numbers of children operated day and night, in six-hour shifts, the workday lasting twelve hours. In those factories that operated only during the daytime, the workday lasted from twelve and a half to fourteen hours, excluding breaks for breakfast and lunch. In some enterprises, like the bast-matting mills, the workday lasted for eighteen hours.[76] Some businesses operated day and night in three eight-hour shifts. Under this arrangement, workers worked eight hours a day during one week and sixteen hours a day during the next one.[77]

Most contemporary observers noted that there was no difference in working hours for children and adult workers. For example, according to an 1871 Moscow city governor report, children usually worked "the same amount of time as adults."[78] Medical doctor F. F. Erisman, an observer of Moscow industries, noted the same tendency during 1879–80. In his reports, he pointed out that children worked the same amount of time as adult workers, from twelve to sixteen hours, depending on the factory's labor organization. Nighttime work was typical for children. In textile mills, children usually worked two six-hour shifts a day.[79] Thus, in most factories the workday for children lasted about twelve hours, and in some enterprises it approached sixteen and even eighteen hours (the same as for adult workers). With the introduction of labor protection laws by the end of the nineteenth century, the workday began to decrease, approaching eleven and a half hours for adults and eight hours for children (see chapter 4).

Most contemporaries noted that although juveniles worked the same number of hours as adults and often performed the same volume of work, they were paid significantly lower rates. The Ministry of Finance commission maintained that the extensive employment of children in industries "leads to a dramatic reduction of wages of children, who work for almost nothing."[80] According to E. N. Andreev, children received from one to twenty rubles a month depending on their age, gender, work, arrangements with the employer, and location. In those cases in which children were provided by their employers with room and board, their wages were significantly lower than those where they subsisted on their own. In cotton-spinning mills, for instance, children worked about twelve hours a day and earned from three to twenty rubles a month—that is, three to five rubles with room and board and up to twenty without these subsidies.[81] The highest monthly wage children received was recorded in a Moscow woolen-cloth mill (up to twenty-five rubles) and in a Moscow tobacco factory (thirty rubles).

Although in a few cases children could make the higher salaries mentioned, the average monthly wages for children were low. Moreover, workers' and children's wages were frequently reduced by various fines and dues for damaged products or broken instruments or tools; for being late to or absent from work; and for the maintenance of certain factory services for workers, such as factory physicians, baths, and so on. As calculated by Dr. Dement'ev for Moscow province, for children under fifteen years of age, the average wage was 2.43 rubles (17.3 percent of the average monthly wage for an adult worker); for juveniles between fifteen and seventeen, the average was 3.35 rubles.[82] The average figures on children's wages from Moscow province roughly correspond to empirewide norms. As mentioned earlier, some contemporaries estimated the cost of basic foodstuffs at about 4 or 6 rubles a month. In Serpukhov (Moscow province), one boy aged seventeen reported to a factory inspector that he received 31 kopecks a day (about 7.75 rubles a month) and spent monthly about 4 rubles on food. Wages of other fourteen-year-old boys were even lower, from 4 to 6 rubles.[83] Hence, the wages of children were often hardly enough to buy food. According to factory inspectors, "the labor of children below 14 years of age hardly pays for their subsistence. Their wages are a more or less significant contribution to their families, but their competition reduces the price of labor."[84]

Children who worked for their fathers or relatives usually received no wages at all. In such cases, children's labor input contributed to the productivity of those for whom they worked and, in turn, increased *their* wages. Children employed by nonrelatives were usually paid an average wage of 2 or 3 rubles a month, with room and board. As described by factory inspectors, the conditions for children hired by adult workers were often grim, especially for those who depended on room and board from the workers who hired them. In most instances, inspectors found that such children not only received no salary but usually owed their employers sums of money for the room and board they had received. Inspector Peskov remarked that the food-supply registers he had examined consistently revealed children's indebtedness to their foremen. "After all," Peskov wrote, "one can [only] imagine the subsistence level of those who have been hired by individual workers."[85] Of course, most factory children lived not on their own, but with their families, and their wages often contributed to family budgets, in which cases the families provided support for the children.

Depending on the number of dependents, an average working family spent a significant portion of its income on food, board, and other necessities. Naturally, most working families lived quite modestly. Depending on their personal circumstances, workers dwelled with other workers in living quarters provided by their factories, rented beds or rooms, or in a few instances owned their own spaces. The first situation was especially characteristic of single workers. Married workers usually rented rooms or quarters. The diet of typical workers in St. Petersburg, as described by a contemporary, usually consisted of rye bread with salt and water for breakfast; for dinner they had Russian cabbage soup (*shchi*) with no more than a half pound of beef (or sometimes no meat) and boiled buckwheat. In the summer, this diet might be complemented with fresh vegetables, such as green onions and cucumbers. This contemporary noted that the diet of a single worker living with a group of workers (*artel'*) was usually more modest that that of his counterparts in a working family. If a single worker drank tea only on weekends, a married one living with a family "gratified himself [daily] with tea and coffee."[86] Late nineteenth-century reports of factory inspectors align themselves with these observations and suggest similarities in workers' diets throughout European Russia. For example, Peskov, who supervised workers' labor and living conditions in private businesses in the Vladimir industrial district in the early 1880s, noted that typical products bought by workers in food markets were cabbage, rye flour, buckwheat, vegetable oil, lard, and tea. Workers consumed meat and fish only on special occasions. It is interesting to note that workers' provisions records sometimes reveal quite high expenses for pepper, which, factories inspectors suggested, in fact may instead represent expenses associated with alcohol, on occasion consumed by children as well.[87] Regarding nutrition, the situation for

Russian workers fit a predominant model. According to scholars of Europe, in the late nineteenth century, "workers almost everywhere [in Europe] remained chronically undernourished."[88]

In many cases, workers bought their foodstuffs from factory stores where prices were generally 5 to 20 percent higher than at regular markets. Only a few businesses, usually large ones, provided their workers with food below market prices. For example, the factory stores of Morozov's cotton mills and Baranov's Troitsko-Alexandrovskaia mill bought food provisions wholesale, directly from producers. In addition, the Morozov enterprises had their own agricultural and livestock farms that supplied the stores of the Morozov factories. In the 1870s, the Morozov enterprises employed many thousands of workers. One Morozov cotton mill in Tver' province used the labor of 4,536 workers, over 16 percent of whom were children under fifteen.[89] Presumably, these workers, including children, did not suffer from malnutrition.

How did industrialization affect the conditions of labor for children? This has been a highly politicized question, which, as a result, has produced controversial but quite simplistic responses. In reality, this question is more difficult to answer precisely than it might appear. Not surprisingly, Soviet scholars of labor have insisted that labor conditions in capitalist factories were extremely poor and even oppressive. Late imperial scholars and factory inspectors, although they often focused on negative and sensational cases, suggested that working conditions were varied and depended on individual factories. Comparing small handicraft enterprises with large mechanized ones, some late imperial scholars suggested that human conditions were better in the former. For example, in his famous "The Factory: What Does It Give to People and What Does It Take from Them," the late imperial scholar of labor Dement'ev wrote that "in small businesses the worker enjoys greater freedom than in large ones that use mechanized technology. [In the latter] things are different. The worker is squeezed into an iron frame. He depends so much on the machine that his own will and emotions are completely suppressed.... Moreover, in small enterprises work usually is conducted during daytime, whereas in large mechanized ones work continues day and night."

Dement'ev pointed out that during the night, small factories could naturally refresh the air inside their workrooms, whereas in large factories that operated day and night and that had inadequate air circulation, workers breathed in stuffy and "unhealthy" air.[90]

Dement'ev's observation about the high degree of workers' dependence on machines in mechanized mills is probably indeed to the point. Still, his somewhat romantic view of traditional workshops hardly finds support in other contemporary descriptions of conditions in small operations. Indeed, labor conditions in many traditional workshops, as described by factory inspectors, were terrible. For example, in the above-mentioned bast-matting shops, according to contemporaries, conditions for workers were the very worst of all described. The factory room where the team worked and produced matts also frequently served as workers' living quarters. All adults and children slept together next to their workplace. Sometimes the same rooms housed domestic animals, such as chickens and pigs. Work rooms were saturated with the rotten smell of wet bast and animal urine and lacked any air circulation. In these same rooms, workers lived, slept, and took their food. One vivid observer of living conditions in these enterprises remarked, "Sometimes a chicken would come up to sleeping children and peck a cockroach creeping across child's face ... but the child's dream is not interrupted." In some bast-matting mills, working rooms accommodated several families. An account of workers' conditions in one such factory in Nizhnii Novgorod (Volga region) stated that "each workshop had 20 bast-matting frames. Each working family sleeps by the frame where they work. There is no other place for sleeping."[91]

In small workshops, children often worked with hazardous chemicals. For example, in the matchmaking industry, in which children under twelve years of age made up about half of the industry's workers, conditions were outright perilous. In 1845, a police report described the work of children in the Shvederskii Match-Making Mills in the Yauza district of Moscow. The mills employed sixty-seven children between ages ten and fifteen. The children worked in low-ceilinged stone rooms, which had neither windows for air nor fans. In these rooms, children dipped matches in sulfur. Consequently, sulfur was boiling in an open tray in the same room during the entire workday. The report noted that "the fireplace is really not quite appropriate for th[e] task [of ventilation]. Although the tray has above it an iron cowl with a ventilation pipe that leads outside in order to draw out the sulfur evaporations, the largest proportion remains in the rooms. Several hundred matches were fixed in a plate and dipped in sulfur. Afterward, the remaining sulfur was shaken off on another tray, where a lot of evaporation also occurred."

According to some accounts, match-making factories sometimes used the labor of four- and five-year-old children. The workday started between five and six in the morning and lasted until eleven P.M., with breaks for lunch and rest. Children received about 1.5 kopecks per one hundred matches. Children employed in these enterprises usually came from extremely impoverished or alcoholic families.[92] Worth noting is the fact that by 1880 the labor in most match-making enterprises remained much the same as it had been in 1845. Of course, labor conditions such as those in the bast-matting and match-making mills were not a novelty and were hardly products of industrialization. During industrialization, however, increased demand for and production of bast mats and matches heightened the demand for child workers forced to labor in such conditions. Nevertheless, bast-matting and match-making enterprises were probably exceptions, and their conditions should not obscure those of other workshops that provided their workers with better environments.

Poor ventilation, inadequate air circulation, and the lack of space between machines, according to most contemporary accounts, were serious issues in most factories. Only some factories that happened to be located in new buildings with wide working rooms had adequate air flow and ventilation. According to Peskov's 1883 account of Vladimir province enterprises, most provincial textile-mill spinning rooms that he visited were wide, filled with light, and had relatively little dust. Machines were installed with considerable space between them.[93]

Nonetheless, working conditions in many factories were undoubtedly bad. Calico-printing factories in St. Petersburg, for instance, astonished some contemporary observers with their "particularly bad construction, as though workers' health [was] absolutely forgotten." St. Petersburg leather-tanning mills stood out as "astonishingly dirty, stench-ridden, and cramped."[94] Similar accounts came from other industrial areas. According to Peskov's description, the preparation rooms in cotton-spinning mills, such as scutching and carding rooms, were usually unsatisfactory. Machines were often set up close to one another with narrow passes between them. Moving parts of machines were in most instances not secured. These machines were often of old construction, their moving and shifting parts insufficiently covered. The moving belts of carding machines were covered only underneath, whereas upper vertical and horizontal belts remained completely unprotected. Dust-removing devices were not always installed.[95]

Almost all the large factories that Peskov inspected had posted instructions about work safety. The only exceptions were small businesses, where such instructions were often absent. One way or the other, factory owners and workers themselves rarely followed safety instructions. Peskov remarked that he never saw workers actually stopping machines while cleaning them, although the rules disallowed cleaning machines while they were operating. Most workers received piece-rate wages and therefore had no motivation to stop their machines for cleaning and never did so. In addition, stopping spinning machines usually degraded the quality of yarn, for which workers were subject to fines. Therefore, cleaning was performed by children who assisted workers while the machines were in full movement. Children were constantly maneuvering around and under the machines in order to clean them. Consequently, at every moment, they risked serious injury.[96]

How did working children view their experience in industries? Evidence regarding children's perceptions of their factory labor remains scarce, most working children leaving behind no account. Their voices are for the most part lost to history. Still, a harshly negative view of their factory employment can be suggested by many children's strong desire to return home, as recorded in police reports about work-related accidents in factories. Indeed, in most instances, children did not want to continue their work in factories.[97]

The Impact of Factory Labor on Children's Health

How did the environment of mechanized factories affect children's health? In an early 1845 report to the Moscow military governor, the chief of the city police wrote about the perilous conditions facing children who worked in one match factory:

> With regard to the health of the boys, they all have poor face color and continuously cough. This happens because the children work in low-ceilinged rooms, under low stone arches, which have neither air holes nor ventilation. In these rooms they cover matches with sulfur and phosphor, which, during the entire day, are melted on a hearth.... During their work, the children are breathing in that hazardous miasma; all, without exception, have a pale, exhausted, and sick look, and constantly cough. If they continue to stay... in such an environment ... they will develop illnesses that will prematurely end their lives.[98]

Doubtless, the exhausting industrial environment and long work hours had a horrendous effect on the health of working children, whose physical development was not complete. Industrial labor led to the physical decline of many factory children. Unlike work in traditional agriculture and cottage industries, labor in the new mechanized factories subjected children to the rapid pace of machinery and exposed them to dangerous moving belts, shifting parts, intense heat, high noise levels, and hazardous conditions associated with dust and the use of toxic chemical solutions.

Of course, in the countryside, children also might work with hazardous equipment, be involved in accidents, and suffer injuries. Still, labor conditions for children in the countryside were much safer. Russian agriculture and cottage industries had long relied on machinery and technology that required a great deal of manual labor. Moreover, as noted earlier, children worked under the supervision of their parents or other adult family members and were assigned work "suitable to their strength." The new factory environment and working conditions, as well as the absence in many cases of parental supervision, exposed children to increased risk that in fact resulted in sickness, work-related injuries, and even death at rates much higher than in agriculture and cottage industries.

It is important to note that general death rates in the countryside were somewhat higher than in cities. Most observers suggested that this resulted from unsatisfactory living conditions in the countryside. Some commentators on childhood have remarked that because the child mortality rate was higher in the countryside than in cities, living conditions for rural children were worse than for urban ones.[99] This, however, should not be attributed exclusively to poor living conditions in the rural environment. Contemporary observers of rural life pointed out that many factory workers who came from the countryside maintained their ties with the village and, after working in factories for a number of years, returned to the countryside, where after a few years they died of natural causes. In this view, their subsequent deaths in the village in all likelihood reflected not so much poor living conditions in the countryside as the harsh factory labor and the diseases and other disabilities that workers developed while working in factories. Therefore, ironically, factory labor was in part responsible for increasing the mortality rate in the countryside. This pertained not only to working adults but to children as well. The data suggests that many rural children who seriously damaged their health while working in factories also returned to the countryside as their strength failed, an undoubted tragedy associated with industrialization.

Observers noted and commented on this sad phenomenon. Dement'ev asserted that the high mortality rate among the rural population resulted from an influx of ailing persons arriving from the cities. He maintained that "one can find statisticians who, on the basis of firm numbers, point out high mortality in the countryside and low [mortality in urban] centers. But only our local *zemstvo* physicians, who maintain medical records for every rural family, know that the real reason for the high mortality is factory [labor]. They know that immediately after they return to the countryside [workers] come to medical establishments with all the signs of incurable lung problems, and in a very short time their medical records would mark them as 'dead of consumption.'"[100] Late nineteenth-century Russian literature also illustrates the tendency of workers whose health was failing to return home to the village. In his short story "Muzhiki" ("Peasants," 1897), about an ill Moscow worker who had just come back to his native village, Anton Chekhov, always a careful observer of

Russian society, pointed out that "even if you are sick it feels better at home and life is cheaper; and it is not for nothing people say that 'home walls help.'" Chekhov's fictional worker dies a few months after his return, a frequent real-life phenomenon as well.[101]

Poor labor conditions, unprotected moving parts, hazardous chemicals, and air rendered harmful by the evaporation of dangerous substances also damaged the health of adult workers, but this harsh industrial environment harmed children's health even more. Perhaps the most revealing remark about the impact of factory labor on children's health was made by a factory manager. In his study of factory labor, Balabanov cited an account written by a correspondent for the newspaper *Russkie Vedomosti*, who observed children working in the dyeing rooms of a cotton factory, where the temperature reached forty-five to fifty degrees centigrade. The children worked on dryers. Startled by the unbearable working conditions in this factory, the correspondent asked the factory's manager what sort of persons these children became when they grew up: "After thinking a while, the manager responded, 'God knows what happens to them. We don't see them at all afterward... They simply perish, totally perish.'"[102]

Clearly, during the period under discussion, the industrial environment exposed children to more harmful conditions than in the countryside. Numerous medical records and accounts point out that children in cotton mills suffered from "an alarming array" of health problems. According to a report received by the Commission for Technical Education in 1874, "In cotton-spinning factories children suffer from anemia. The hands of children who clean machinery are irritated with a rash because of mineral oil. Children who work in preparatory shops suffer from soreness of the breathing canals and throat."[103] According to factory inspectors, in some cotton factories, children employed in preparatory shops were "dirty in the extreme, covered with some kind of odd lesions, and looked very exhausted."[104]

In addition, for technological reasons, textile mills maintained very high internal temperatures. For example, spinning rooms maintained high temperatures and humidity to help reduce the breaking of threads.

But high temperatures exhausted workers. One factory worker, A. A. Voskoboinikov, wrote in the social and political journal *Biblioteka dlia chteniia* ("Reading Library") in 1862 that children who worked in the printing rooms of a calico factory, where the temperature reached forty degrees centigrade, had "yellow faces, red, swollen eyelids, an unhealthy look, and hollow chests. This is the indisputable evidence and inevitable consequence of some two to three years of employment in cotton factories." Voskoboinikov claimed that labor in cotton factories "prevents the physical development of children."[105] These examples illustrate the impact of working conditions in the textile industry, which, of course, employed the largest number of children. One may assume that numerous children who worked in textile factories seriously damaged their health and that many may have perished.

As during the 1840s, conditions for working children remained the most harmful in the match-making industry. According to the medical records of children who worked in this industry, their skin was pale, flabby, and dry. They had face and leg edema, dry and spotty tongues, a weak and irregular pulse, shortness of breath, and a dry cough. In one case, in a Moscow match mill, an accident poisoned eleven children with sulfur fumes. They were sent to a hospital, where three soon died of hemorrhaging of the brain, pneumonia, and typhus fever brought on by the exposure.[106] An autopsy commissioned by the local authorities for one of the boys, fifteen-year-old Sergei Safonov, showed that his stomach was empty; his lung tissue was feeble, flabby, and covered with many purulent tumors; and both his heart ventricles held coagulated blood. The postmortem skull examination revealed that the cerebrum had been overfilled with blood. The examining pathologist diagnosed hemorrhage of the brain caused by consumption.[107] After an investigation, the Moscow authorities prohibited the mill from the practice of apprenticeship and from employing individuals under age eighteen; they also required the owner to isolate the match-making room from other areas of the mill and install ventilation cowls.[108]

Labor conditions for children seem to have been no better in some other industries. In a sugar plant, for instance, as described by a contemporary, "children of eight to ten years of age and sometimes even seven years old scaled boilers in extremely harmful conditions.... [The children] suffocate from the dust and soot."[109] In many factories, the absence of proper air circulation led to poisoning by hazardous chemicals such as chlorous oxide and sulfur.[110] Children who worked with hazardous chemicals, according to medical reports, suffered from serious lung problems. *Cachectic* and *pale* were the terms most contemporaries used to describe child factory workers. They noted that "the dusty and asphyxiating atmosphere of the factory" was "harmful for the child's immature organism."[111]

In addition to numerous general illnesses brought about by the new industrial environment, children were also disproportionately subject to work-related injuries. The absence of proper air circulation, the cramped spaces, and the lack of covering over moving and shifting parts often led to work-related accidents. If official investigations of work-related injuries occurred, information about them is difficult to locate, although, as we shall see, statistical data tell the story nonetheless. The following document, one of very few surviving pieces of evidence produced by children themselves, illustrates the problem. In 1857, a work-related accident happened to a sixteen-year-old boy, Andrei Agapov, who was employed at a wool mill owned by the merchant Nosov in the Lefortovo district of Moscow. In a police report Agapov testified:

By faith I am Orthodox Christian, take Holy Communion every year. I know literacy, but because of the disorder of my right hand, on which the fingers were injured, I cannot affix my signature. I have been living at the mill of the merchant Nosov since the autumn of 1856 [and work] as a helping boy on the shearing machine. Last March, the 23rd, right before breakfast, when I was on duty with my fellow worker Nikifor Nikiforov, I tried to straighten the cloth when it began to jam.... Two fingers of my right hand went with the cloth on the knives that cut nap. These knives cut off the nail to the bone on my middle finger and cut off flesh to the bone on the fourth one. I had pulled out my hand and was so scared that I did not feel any pain until the local physician arrived and dressed the wounds. I was immediately sent to a hospital.... Meanwhile, I am feeling all right and ask to be sent back home.... This was an accident, and I do not blame anyone else for doing this deliberately.[112]

The police investigation of Agapov's case indicated that the boy had worked under the supervision of an overseer, an eighteen-year-old worker. Although the police found that Agapov was himself responsible for this accident because of "his own carelessness," the employer compensated the boy with a sum of money. After his recovery in the hospital for workers, he returned to his home village.[113] Of interest is that the overseer in this particular mill, an eighteen-year-old, may himself have required adult supervision. The important question of children's and juveniles' mental and neurological capacities will receive attention below.

Similar cases were recorded in other enterprises, suggesting that accidents that involved the loss of or damage to extremities were common. For example, in the Moscow cotton mill owned by the merchant D. F. Fink, in the city's Khamovniki district, a seventeen-year-old boy, Peter Ivanov, lost two fingers of his right hand while working with a horse-powered belt drive. (It is noteworthy that in the legal records seventeen year-old Ivanov is defined as a boy [*mal'chik*] a designation that reflects state perceptions of adolescence.) Ivanov requested a full 2,000 paper rubles for his loss but received compensation of 100 paper rubles from the merchant and 250 paper rubles from the state, respectable sums by standards of the time. After recovery, he returned home, as he strongly desired.[114] In most such cases, the reports underscored the children's "own carelessness and neglect" and held them responsible for the accidents. In the abovementioned Agapov case, the official report claimed that the incident happened because of a "prank" and "unintentional imprudence."[115] In all recorded incidents, children were given appropriate medical care at the site or sent to a medical establishment. In most recorded cases, the injured children expressed a desire to return home, as was the case with Agapov and Ivanov. The reports usually state that the employers paid some benefit or other, which seems to have reflected a certain paternalism. Although there is nothing implausible about the finding that the children themselves were responsible, in a certain sense, the question of ultimate responsibility remains open. Regardless, it should be emphasized that all work-related accidents, as well as work disputes, fell under the auspices of the local offices of the Ministry of the

Interior. It was these very officials who first observed such cases and first urged the need for legislative measures (see chapter 3).

All evidence suggests that children were more prone to work-related injuries than adult workers. As noted, the most coherent data on working conditions for children comes from the 1870s and early 1880s. The Sokolovskaia Cotton Mill (Vladimir province) left behind valuable information on injuries associated with factory employment. During 1881–82, of 165 registered accidents, 87 (53.0 percent) occurred among working children, whereas children accounted for only 10.8 percent of the factory labor force. The number of registered accidents indicates that, in the given period, about 16.0 percent of children employed at the mill experienced accidents, as opposed to only 2.7 percent of adult workers. Most accidents involved cuts; wounds; broken limbs; and fractured arms, fingers, and legs, which often led to amputation.[116] The most frequent accidents happened among children who pieced thread and set up bobbins. The latter task, performed mostly by male children (77.7 percent), was the most dangerous operation. About 37.0 percent of accidents in the spinning shop were associated with setting up bobbins.

Numerous other accounts support this data from the Sokolovskaia Cotton Mill. Medical and police reports, for instance, illustrate that the most common work-related accidents involved hand and limb injuries.[117] According to Moscow government officials, a similar pattern of child injuries existed in Moscow and its province. They confirmed that children were more vulnerable to injuries than adult workers.[118] A St. Petersburg government factory commission set up in 1859 also reported that the highest number of work-related accidents were suffered by children. The commission found that during a certain period, cotton factories experienced forty-eight accidents with "serious consequences" that required a physician's attention among children age fourteen and under and twenty-eight accidents among children between ages fifteen and sixteen, whereas seventy-two accidents involved adult workers. The number of accidents among children and juveniles (seventy-six) exceeded the number among adult workers, although the overall number of working adults greatly exceeded the number of employed children.[119] Work-related accidents also caused deaths. For example, according to police records, in the Guk factory in St. Petersburg, two to four children died annually.[120]

Most contemporaries attributed the ill health of factory children to their physical immaturity and to hazardous labor conditions in factories. They suggested that labor in mechanized factories and the fast pace of new machines required greater energy from working children than they possessed. Voskoboinikov pointed out that "the labor burden on children who work on mule machines exhausts them." The highest number of injuries occurred among workers who worked on these machines. Overall, work-related accidents reached their maximum level in textile (cotton-spinning) factories and metallurgical plants, both industries with a high level of mechanization.[121]

Recent medical and pediatric studies help explain the high rates of work-related accidents among working children. These studies shed significant light on the differences between adults and children as regards physical condition and mental abilities. For example, one study finds that the eye movements of preschool children differ from the eye movements of adults, a factor that limits children's ability to acquire adequate visual information.[122] One may imagine the significant impact this factor would have had on the labor of young children and ultimately on the higher number of work-related accidents among them. This was especially the case when children were working with high-speed mechanized equipment. Alongside the impact of incomplete physical development, another possible explanation for the high rate of work-related accidents among children can be found in recent research about neurology and developmental psychology. This research emphasizes the different stages of development of the human brain in adults and children, producing different patterns of behavior and responses to the environment. Children cannot think in abstract terms.[123] Thus, children's behavior and responses to the factory environment and machines were dissimilar from those of adult workers. Below a certain age, children and young adults fail to project the consequences of their actions. In the case described above, the eighteen-year-old overseer himself likely did not have the necessary mental attributes to work safely and could hardly provide appropriate supervision for younger workers. Medical research suggests that humans reach complete cerebral maturity only by age twenty-one. In turn, brain immaturity and low ability to coordinate attention would have increased the incidence of work-related injuries among working children.

Although medical knowledge about the brain and physical capacities was limited at the time, various publications of the 1860s and 1870s nevertheless emphasized the need to deal with the issue of child labor. An 1875 editorial in *Vestnik Evropy* pointed out that data on child industrial labor was sufficient to promote a legislative effort. "Every passing year," claimed the editor, "threatens the health and even lives of numerous factory children, poor victims of need."[124] In his 1871 report, the Moscow city governor maintained that "the young generation is declining physically" because of exhausting work in factories.[125] In 1878, the Moscow city governor called for energetic legislative measures to cope with industrial injuries among children.[126] Some contemporaries even identified the death rate and declining health among young factory workers with warfare. "The bloodiest wars," wrote an observer in 1882, "seem innocent jokes ... if compared to these losses of life and health [in industries]."[127] This bitter, even exaggerated expression reflected growing concern among many statesmen and public activists about the decline in the health of the younger generation and its potential consequences for the security and well-being of the empire. Many contemporaries realized that the factory was not a good place for children. These concerns contributed to the emergence of attitudes opposed to child labor and to appeals for child labor protection laws from state officials, public figures, and intellectuals.

A lesson in mowing, early 1900s.

A lesson in shoemaking, 1890.

Working at a plant, 1900.

Haymaking with a mowing machine, 1900.

Peasant children, early 1900s.

A weaving shop, 1900s.

3 ✹ Public Debates and Legislative Efforts

AS NOTED EARLIER (see chapter 1), during the early nineteenth century, most state officials perceived child labor as a normal practice essential for the upbringing and education of children. Prominent statesmen and public figures, such as N. S. Mordvinov and P. S. Nakhimov, viewed child labor as morally justified and useful. During the 1860s, however, such attitudes began to languish and gradually gave way to voices that opposed child industrial labor. Unfavorable information about the impact of the new factory environment on children's health induced contemporary commentators to question the moral aspects of employing children in industries. Many state officials and public figures began to doubt that the factory was an appropriate place for children's apprenticeship and work. From initial approval, their attitudes shifted toward emphasizing the need for restricting the employment and labor of children.

The appeal for child labor protection laws initiated by state and local bureaucrats produced an important public discussion of child industrial labor among state officials, industrialists, academicians, and all others concerned about the issue. During the 1860s and 1870s, the government organized several commissions whose purpose was to inspect labor conditions, review existing factory legislation, work out new factory labor regulations, and promote discussion of these regulations. Although the impetus for this discussion usually came from local and imperial government officials, during the 1870s, it also involved much broader segments of society, including academics, medical circles, and industrial and other public associations, as well as journals and newspapers. The debates about child labor helped form new perceptions of children's industrial employment and education.

What impact did the debates have on general perceptions of childhood? How did the debates change the attitude of state officials about child labor? What impact did all of this have on actual legislation about child labor? The answers to these questions are important in and of themselves, in no small measure because existing historiography has not even raised, much less exhausted, these issues. The whole matter takes on added significance, however, because it opens up an entirely new perspective on late tsarist lawmaking and governance, challenging certain long-dominant historiographical interpretations.

Early Legislative Proposals and the Discussion of Child Labor

By the late 1850s, government officials recognized that the existing labor regulations for private businesses—the 1835 and 1845 decrees—no longer suited contemporary needs.[1] This in essence signified the beginning of the labor question in Russia, as mentioned in Reginald E. Zelnik's study of the early labor movement. Still, Zelnik and other scholars link the issue of early labor laws primarily to Russia's defeat in the Crimean War. This study adheres more closely to the preoccupations of state and society as expressed in the documents that emerged from the discussions and from the commissions appointed to formulate new factory regulations. These documents in fact mainly reflect concern about working conditions in factories and especially emphasize the negative impact of industrial labor on children.[2] One governmental report admitted that "the frequency of work-related accidents among workers, and especially working children, requires new regulations" for factory labor.[3] In 1859, the imperial government set up two commissions, one under the Ministry of Finance to review the Factory and Apprenticeship Code, and a second under the St. Petersburg governor to "thoroughly investigate" working conditions in the city's private factories and workshops and formulate new employment and labor statutes for St. Petersburg.[4] Both these commissions were headed by A. F. Shtakel'berg, an expert on legal issues regarding factories and workshops in Russia and Europe.[5] Both commissions included local and imperial government officials, public figures, physicians, educators, and a few business representatives. The appointment of these commissions signified the beginning of a process of labor-related legislation and debates in Russia. Without exaggeration, one might say that this launched the Russian labor question. Local offices of the Ministry of the Interior and, in particular, its district medical and police departments were usually the primary institutions to consider local labor issues. On an ongoing basis, they settled labor conflicts and dealt with work-related accidents in private industries. Therefore, it was not accidental that the initiative for studying labor conditions and introducing labor protection laws came from these concerned local bureaucrats. When the St. Petersburg commission examined working and living conditions in the city's industries, it confirmed that the factory was an unsafe place for young children. According to its report, "factory work and ... the stuffy and dusty [factory] environment have a fatal impact on children's immature bodies.... [Factories] overwork children and treat them harshly and cruelly." The commission maintained that the state should "protect the younger generation from being subjected to exhausting factory labor." It suggested the strong need for restricting child labor in the city's industries.[6] The ministries of Interior and Finance asserted that child labor regulations should not be limited to the capital but introduced in other industrial areas of the empire as well.[7]

The two Shtakel'berg commissions addressed multiple aspects of industrial labor and proposed quite similar measures for restricting employment and labor for children in industries and domestic services. The commissions suggested that the employment and apprenticeship of children under the age of twelve should be prohibited entirely. Children age ten to twelve could take an apprenticeship only when they were apprenticed by their own parents or, in the case of orphans, by close relatives who served as their guardians. Following the language of earlier laws, the commissions specified that in these cases children under age twelve "should be assigned tasks according to their physical abilities." The St. Petersburg governor's commission proposed to limit the workday for children and juveniles age twelve to fourteen to twelve hours, including a two-hour break for lunch and rest, and suggested that the workday for children under sixteen should be only between five A.M. and eight P.M. Thus, the St. Petersburg commission suggested the ten-hour workday for children age twelve to fourteen and a ban on nighttime work for children under sixteen. Later, in 1862, the Finance Ministry's commission went even further and proposed the ten-hour workday limit and a ban on nighttime work for all children and juveniles under age eighteen.[8]

Regarding the education of child workers, the commissions came up with several rather vague ideas. They proposed that factory owners should be responsible for the general intellectual development of working children. Businesses should not prevent working children from attending Sunday and evening schools. Sizable businesses with a large number of workers should found their own basic literacy schools for their workers.[9] Still, most of these legislative proposals lacked specificity and appear to have been advisory rather than obligatory in their formulation. Even so, their significance lies elsewhere: They represented the very first foray of the government—and, for that matter, the public realm—into questions about working children's general welfare.

Both Shtakel'berg commissions also addressed working and social conditions for workers, adults as well as children. The commissions were concerned about work safety in industries and suggested that factory owners be required to provide their enterprises with safety measures, such as shielding moving parts of machines and providing proper air circulation and lighting in workshops. Owners of enterprises were to be responsible for formulating work-safety instructions and posting these instructions in places accessible to all workers. The proposals obliged owners and managers to inform workers about potential dangers that work and machines could pose to workers' health. In other words, owners could not employ workers without informing them of potential risks and safety rules. In their turn, workers were supposed to learn work-safety rules.[10] Additionally, the commissions' proposals specified the financial compensation of workers for work-related accidents and sickness during the period of their disability. These provisions would oblige owners in these cases to pay all medical expenses, including those for physicians and medicines.[11]

Furthermore, the Finance Ministry proposal included provisions that provided a legal basis for workers to create their independent mutual assistance associations, such as *zemliachestva* (fraternities) and *arteli* (cooperative work groups), which at the time existed on an extralegal basis.[12] The proposal also contained provisions on business arbitration courts (*promyshlennye sudy*) where workers and employers would be equally represented, responsible for mediating and containing conflicts between employers and workers.[13] In order to implement and supervise laws over factories, the commissions suggested the introduction of state-paid factory inspectors and the imposition of penalties on those who evaded the regulations. The penalties would include fines of ten to three hundred rubles and, in some cases, specified administrative sanctions.[14]

Apparently, these commissions, headed by Shtakel'berg, an individual who had studied foreign labor laws, took into account labor legislation in other European countries. In fact, the commissions thoroughly examined contemporary Western European labor laws, as well as existing legislation that already regulated some Russian state and private industries. Nevertheless, in many respects, the commissions' propositions to prohibit children below twelve years of age from employment and to limit the employment of children between the ages of twelve and eighteen in private businesses went far beyond contemporary European legal norms. As noted, several European countries, such as Britain, France, and Prussia, had introduced some limitations on child labor but had banned outright only the employment of children under eight or nine years of age, and that only in certain categories of production. For instance, an 1833 British statute banned the employment of children eight and under only in textile mills that used steam or water power, leaving untouched the numerous concerns that still operated without these new technologies.[15] Thus, regarding the minimum age for employment, the commissions clearly followed their own norms: those that had been established earlier in some Russian industries and traditionally practiced in the countryside.[16]

Regardless, these provisions designed by the St. Petersburg commission for the city's private industries did not for the time being become law.[17] Some Soviet historians of child labor argued that opposition to their enactment came primarily from industrialists in Moscow and other central provinces where "traditionally organized" industries depended heavily on child labor.[18] This assertion, however, is hard to justify. As we have seen (chapter 2), St. Petersburg industrialists also employed many children. In fact, they revealed no less concern about the law's enactment than did entrepreneurs from central provinces where the proposed provisions actually did not apply.[19] St. Petersburg industrialists insisted that the proposed child labor restrictions for the city would place their industries at an obvious disadvantage compared to other industrial areas of Russia where child labor would remain unregulated. They recommended instead nationwide regulation of child labor. This was one of the reasons why the ministries of Finance and the Interior suggested in 1859 that factory regulations should not be limited to St. Petersburg but expanded to all industrial centers and "required for all" private businesses.[20]

Nevertheless, the St. Petersburg commission had at least some positive accomplishments. The commission gathered valuable data on factory labor in the city and its district (*uezd*) during 1859–60 (see chapter 2). In addition, and perhaps more important, some of the commission's suggestions found their place in a new statute on state mines and metallurgical mills, the provisions of which were enacted in March 1861. In June 1862, similar regulations were introduced for private mining and metallurgical enterprises. These and some earlier statutes set the minimum age for employment in these enterprises at twelve years, prohibited underground work for children between the ages of twelve and fifteen, and introduced factory inspectors. The decrees obliged managers of state-owned businesses and owners of private businesses to maintain schools for employed children and for children of their enterprises' workers. The decrees also introduced free medical care for work-related injuries and free basic medical services for workers.[21] In addition, the 1861

"Provisional Rules on Employment for State and Public Work" allowed railroad-building workers to organize workers' associations (*arteli*).[22] In this way, these approaches to labor problems, first publicly articulated by the Petersburg commission, began to work their way into tsarist labor regulations.

The efforts at reform continued in other ways. The Finance Ministry commission continued its legislative effort and soon compiled a new proposal, which was published in 1862 and sent out to provincial governments and industrialists' associations for review, discussion, and suggestions. The new proposal received consideration and provoked a lively discussion in the Manufacturing Council (a corporate association of Russian entrepreneurs and industrialists) and its Moscow Section (Moskovskoe otdelenie manufakturnogo soveta), which included industrialists from all the central industrial provinces of Russia. The provisions that addressed child labor were the most controversial. Although many entrepreneurs agreed that the state should introduce some regulation of child labor in private businesses, the dominant attitude toward the proposed specific restrictions was negative. Some discussants suggested following the examples of France and Prussia in limiting the workday to ten hours only for children under age sixteen, not eighteen, as the proposal maintained. Others asserted that the age to begin employment should be lowered to eleven and that the workday for children age eleven to fifteen should be twelve hours in two six-hour shifts, already the norm in many textile mills. Most, if not all, industrialists rejected the idea of prohibiting nighttime work for children between the ages of twelve and eighteen.[23]

How did industrialists justify their opposition to the proposed legislation, and what were their real reasons for opposing it? Most industrialists argued that any restrictions on the workday and on nighttime work for children would ultimately adversely affect the labor of adult workers who were assisted by children and, in turn, hinder the whole production process. For instance, some insisted that, without the help of children, adults could not conduct nighttime work at all. Other industrialists were concerned that the workday limit and the ban on nighttime work for children would lead to a rise in production costs that would consequently make their businesses unprofitable. They maintained that their factories would need to hire more adult workers to replace children. Most industrialists called for no labor regulations for children age twelve to eighteen.[24] Industrialists expressed their concerns about the enactment of the new child labor law at meetings held in the Manufacturing Council and its Moscow Section and in letters sent to these associations. Perhaps the most active opposition to the proposed legislation came from the entrepreneurs of Russia's central provinces, who were well represented in the Moscow Section, although St. Petersburg entrepreneurs also failed to support many of the draft's provisions. The central provinces, with their traditionally organized businesses and reliance on manual labor, heavily relied on children. For instance, the Khludov brothers, textile entrepreneurs from Tver' province, strongly opposed the ban on nighttime work for children, stating that this provision would eliminate nighttime work for adult workers as well, because "adults cannot work without children's assistance." The Khludovs insisted on a minimum age for employment of eleven years and a thirteen-hour workday for children between ages eleven and fourteen.[25] Likewise, certain Tula textile entrepreneurs claimed that "any restriction on the labor of children age twelve to eighteen was totally unacceptable."[26] Various owners of glass-making works, an industry that employed a high proportion of children, joined together and wrote to the Manufacturing Council that "limitations on child labor would mean the complete destruction of the entire glass-making industry in Russia." The glass makers maintained, probably accurately, that in their industry children were hired by their fathers or other relatives and worked under their supervision. The manufacturers also claimed, perhaps less accurately, that the children performed easy tasks and "earned their bread almost playfully."[27]

Many entrepreneurs questioned the drafts' provisions regarding work safety. They argued that these provisions might give workers an advantage in explaining away work-related incidents based in fact upon their own lack of attention, allow them to shift the blame onto employers, and claim compensation.[28] Entrepreneurs also rejected provisions that provided for workers' associations. The Manufacturing Council's Moscow Section argued that "instead of using this opportunity [to organize themselves] for their own good, ... led by some kind of conspirators, who will immediately arrive, workers will use it for harmful ends."[29] At this point, industrialists were supported by certain Ministry of the Interior officials who also questioned these provisions, arguing that they might stimulate "a spirit of solidarity among the masses, [facilitate] strikes, and finally [encourage] disobedience among the working population."[30] Some entrepreneurs emphasized that the proposed law seemed to show too much concern for workers, most of whom were adult, self-reliant, and responsible. They claimed that the law's provisions deprived both factories and workers of the freedom to negotiate individual work contracts and did not in any way prevent parents from exploiting children at home.[31]

During the council's debates, many employers expressed humanitarian concerns about children's families and welfare, arguing that the law's enactment would indeed serve children badly. For example, the Khludovs, like most employers from other industrial provinces, stated that "children, having lost the opportunity to earn money in factories, would not be able to contribute to their parents' incomes ... and instead of [working] in a light-filled and healthy factory building would damage their health in the stuffy atmosphere of their homes."[32] The manufacturers argued that the proposed restrictions on children's employment would decrease the incomes of workers' families and make it impossible for them to give their children a proper education.[33] Similarly, the Tula entrepreneurs insisted that the new regulations on child labor would have "a negative impact on production and, at the same time, bring no benefit for children because the easy tasks children perform cannot harm their development," whereas children would lose the opportunity to earn some cash and thus support their families.[34] Most employers claimed that children were usually assigned tasks that fitted their gender, age, and physical abilities.

Whether industrialists resisted the proposed restrictions out of genuine benevolent concern for peasant families or simply deployed such arguments as a rhetorical device to justify their opposition is an open question. In any case, their concrete suggestions about the minimum age for work and the length of the workday clearly reflected strong collective entrepreneurial motivations. Their almost unanimous opposition to the nighttime-work ban and the work-safety provisions reflected their desire to preserve the employment of children. As noted earlier, children usually assisted adult workers or performed ancillary tasks that, employers believed, could not be performed by adult workers. Indeed, many children were hired as assistants by foremen, sometimes their own fathers. Thus, the proposed legislation conflicted with the tradition of family labor practiced in many businesses. One of the most serious concerns of business owners was that the elimination of child labor would lead to the closing of many factories. Entrepreneurs, especially those who owned smaller, traditionally organized workshops that employed many children, obviously worried that the replacement of children by adult workers, a more costly workforce, would increase prices for their goods and place them out of reach of the majority of the population.

Nonetheless, not all employers opposed child labor regulations. A few paternalistic or philanthropic voices among industrialists supported these restrictions. According to V. Iu. Gessen, St. Petersburg industrialists gave greater support to child labor regulations than did entrepreneurs in the Russian central provinces. Gessen found that some of the city's industrialists even suggested raising the minimum employment age to thirteen and banning children's employment in the most harmful and hazardous industries.[35] By contrast, the labor historian V. A. Laverychev has noted that St. Petersburg industrialists also viewed provisions regulating employment ages as "disadvantageous."[36] In point of fact, both historians were correct. Some Petersburg industrialists opposed restrictions on child labor, whereas industrialists who owned large mechanized enterprises were most likely to support restrictions on child labor. In all likelihood, this latter position reflected these industrialists' perceived economic advantage. Many St. Petersburg employers supported child labor regulations because businesses in the city were large, mechanized, steam- or water-powered factories, whereas in the central provinces small workshops with manual labor and old production methods predominated. As suggested earlier (see chapter 2), in order to increase output and maintain low production costs, small workshops used child labor more extensively than modernized factories, although the overall numbers of children employed in both remained high. Children sometimes composed about 40 percent of the workforce in small workshops, whereas in mechanized factories they made up from 12 to 21 percent of the workforce. The owners of large mechanized factories doubtless realized that, if enacted, the proposed law would reduce the output of traditional workshops and thus give them a competitive advantage. One historian of child labor in Britain, Clark Nardinelli, describes a similar tendency in early British industries. Owners of large mechanized factories equipped with steam engines were among those who supported the 1833 child labor legislation in Britain, whereas owners of small traditional workshops opposed it.[37]

As the next step in the process, the legislative draft and the industrialists' opinions regarding it went to provincial governments for review. With a few exceptions, most government officials at the state and provincial levels supported child labor protection legislation. Many provincial officials revealed their skepticism about the industrialists' humanitarianism regarding children's welfare. In their reports, provincial governors supported the proposed law. From their perspective, the governors realized that the use of children in factories had increased during past years and required state intervention. As previously mentioned, local governments, especially in the person of their district police and medical offices, were usually the first to be confronted with workers' complaints about working conditions and associated health problems. Many governors became seriously concerned about the growing number of work-related accidents among children. Provincial governors felt that it was the paternalistic obligation of the state and the ruling elite to take care of working children and thus improve the well-being of the empire. Thus, during the 1860s, the discussion about labor laws generally involved two groups: the industrialists on the one hand and state and local provincial officials on the other.

In characterizing the industrialists' voices in opposition to the proposed labor regulations, the Tver' governor Count Baranov expressed doubt that business owners were really concerned about working families' well-being. Questioning the manufacturers' position, he bitterly remarked that they "supported the most unethical practices." The governor continued that "it is known that the industrialists do not think about people's welfare and the education of peasant children but only about their own pockets.... They simply exploit [their workers and] their ... abilities."[38] Baranov's indignant reaction signified a notable change of attitude about child labor that began to take place among government officials. Many state and local bureaucrats were increasingly outraged by the realities of child factory labor. They began to characterize the practice as "unethical," "immoral," or "morally unacceptable," whereas earlier on, we may recall, they had almost universally viewed such labor as normal apprenticeship.

Although the concern employers expressed about children's families may have been aimed at concealing their real reasons for this view (to maintain production and exploit the cheapest labor), their arguments nonetheless reflected the harsh economic realities for many peasant families. The earlier discussion (chapter 2) reminded us that many impoverished rural and urban families under economic duress—especially those with dependent children—had to send their older offspring to factories. The wages children received often made an indispensable contribution to their families' budgets. Furthermore, some contemporaries still doubted that the proposed legislative measures would have any positive impact on children's lives in general. They closely associated children's factory employment with poverty, which would hardly be overcome by the introduction of a restrictive law. Others argued that restrictive measures would not eliminate child labor at home, in agriculture, and in cottage industries, where working conditions were sometimes as harsh as or even worse than in new modernized factories.

The governor of Vladimir province, one of the few provincial governors who remained openly skeptical about the proposed regulations' potential effectiveness, noted that it would be "more humane for children and juveniles to work in factories than to stay at home."[39] He argued that the proposed limits on children's employment in factories would inevitably lead to an increased labor burden on children in agriculture and cottage industries, where, he maintained, working conditions were in many cases worse than in factories and where state control over child labor would be almost impossible. He insisted that "the child's immaturity cannot serve as an adequate basis for limiting his freedom of employment. Because of the increasing population, it would be more beneficial and humane if children and juveniles worked in factories rather than staying at home and becoming a burden for their parents, who ... will send them ... to harder work in small workshops that easily escape government control."[40] The Vladimir province governor's arguments that restrictions on child labor would simply result in a shifting of children from larger factories to smaller workshops and an intensification of their labor in agriculture and cottage industries may have been well-grounded. Still, like some entrepreneurs, he also may have had in mind the welfare of the province's industries. Vladimir province was an important center of Russian textile production, where, it so happened, small traditionally organized workshops still prevailed. These establishments relied heavily on the labor of the local peasant population, including many children.

In 1865, discussion of the legislative proposal and the provincial governors' opinions about it returned to the Council of Industrialists, where negative opinions about the proposed restrictions still predominated. Most industrialists continued to express their doubts, arguing that the restrictions "would neither do any good for industries, on the one hand, nor bring any benefit to children on the other." They continued to maintain that the limitations on the workday and the ban on nighttime work for children under age eighteen could have harmful implications for industry, as well as for the children and their families.[41]

Entrepreneurs and others who opposed the law clearly strived to develop discursive strategies to justify their opposition. To reinforce their arguments and make them sound more dramatic, some entrepreneurs even stated that the enactment of child labor regulations would hamper the entire industrial development of Russia.[42] Some employers coupled arguments about the law's negative implications for the nation's economy with expressions of concern for the well-being of working families. The most common strategy involved expressing two arguments concurrently. In its official opinion sent to the Finance Ministry's commission, the council recommended that the minimum age for employment be lowered to eleven years of age and that the ban on nighttime work and the limit on the workday apply only to children between ages eleven and fifteen.[43]

Regardless, industrialists and the Finance Ministry proved unable to reach a compromise. Despite industrialists' strong opposition to the proposed regulations on child labor, and despite the concerns expressed by some statesmen about the regulations' potential ineffectiveness, the commission insisted on their enactment. Its members believed that most industrialists' concerns were either illusionary or highly exaggerated. To counter their arguments, some ministry officials argued that even if the restrictions on child labor led to some increases in production costs and reductions in profit, such regulations, in general, would benefit the nation's economy as a whole. They maintained that "if consumers would pay a little higher price for goods, these prices would be based on more adequate labor conditions and, furthermore, society would not lose the entire generation of children who today are subjected to factory labor." The commission believed that new regulations were crucial in protecting the younger generation from exploitative and abusive labor in industry. Consequently, the commission found it impossible to take into account the industrialists' arguments and accommodate their suggestions. All the proposed provisions remained unchanged.[44]

In 1866, the commission sent its legislative proposal to the Ministry of the Interior for approval. Presumably responding primarily to business and industrial opinion, the ministry gave it no further consideration, and as a consequence it did not become law. The new labor act introduced in 1866 retained most of the old provisions of the 1835 decree. Child labor in private businesses remained unregulated. As regards some other labor-related issues, the discussions and proposals did bear fruit. In 1866, for example, the Imperial Committee of Ministers approved the enactment of employers' liability provisions, which included free medical care for work-related accidents and some paid basic medical services for workers in all industries. (As mentioned earlier, in 1861 and 1862, similar measures had been introduced in the state and private mining industry.) The new provisions obliged all businesses with one hundred or more workers to maintain a medical doctor and keep hospital beds at the rate of one bed for every one hundred workers (ten beds per one thousand). The introduction of these provisions was provoked by the outbreak of the 1865 cholera epidemic. The Moscow governor strongly supported their enactment as a preventive measure against the spread of the disease in Moscow and its province. Unfortunately, despite clear stipulations for the implementation of these rules, most businesses, in the absence of inspectors, evaded the new regulations, as reported in 1885 by newly appointed factory inspectors. Only a few enterprises maintained medical facilities for workers by that year.[45]

Why did the enactment of this legislative proposal, so strongly urged by some elements of the state structure, fail in 1866? Following the lead of V. I. Lenin, Soviet scholars of labor argued that the failure to enact the provisions of the St. Petersburg and Finance Ministry commissions resulted from the provisions' "unrealizable" nature. In this view, the "liberal ideas" that the legislative project embodied could not materialize within the existing autocracy, which had no "serious stimulus" to enact the law.[46] This approach neglects the fact that by definition, "classical liberalism" rejected the idea of state intervention in the economy and labor relations and emphasized instead conceptions of laissez-faire, individualism, and "freedom of contract." In this regard, the failure to adopt a universal regulatory labor law indeed signified state adherence to liberal policy in the matter of labor relations in private industries. After all, the government displayed a distinct readiness to intervene in factory life and serve as an arbiter in labor relations when it wished. By nature, after all, autocracy is interventionist. It was the state and local government that initiated labor legislation and set up the legislative commissions. Moreover, as noted, in 1861–62, the state introduced a new universal labor statute for the mining industry and the "Provisional Rules for Employment on State and Public Work" and, in 1866, the employers' liability act. Some provisions of these acts regarding the education of employed children, employers' liability for work-related accidents, medical assistance for workers, a factory inspectorate, and workers' associations find their counterparts in the commissions' proposals.

A somewhat different explanation for the law's failure may be found in the opinions of certain late nineteenth-century statesmen. They suggested that the proposed restrictions on child labor failed to be enacted because they were embedded within general labor legislation that in turn involved too many "diverse and complicated" aspects, tried to resolve too many issues, and touched upon the interests of too many interest groups. For example, for reasons having to do with its police and control functions, the Ministry of the Interior rejected the draft's provisions regarding workers' associations and labor-dispute arbitration courts and therefore did not support the law as a whole.[47] Holding a different view, the late imperial historian of Russian industry M. I. Tugan-Baranovsky suggested that the staunch and effective opposition of most entrepreneurs helped kill the new labor law.[48] Perhaps a combination of the two versions best explains the result. The industrialists' opposition to child labor regulations, combined with the reluctance of some government officials to adopt certain other labor policies, especially as regards empowering worker self-organization and protest, prevented the passage of the law.

Their occasional paternalism aside, industrialists in general opposed any state intervention in the economy and labor relations. Their resistance to any significant restriction of child labor indeed proved too strong to be overcome at this time. In 1869, the Moscow Section of the Manufacturing Council, under renewed pressure from the Finance Ministry, again discussed the regulation of child labor. Predictably, the industrialists again opposed the law's enactment. The entrepreneur-oriented council still maintained that the minimum employment age of twelve years stated in the proposal was "incompatible with the needs of industries" and suggested instead reducing the minimum employment age to eleven years. The council agreed to limiting the workday for employed children between ages eleven and fifteen to ten hours during the daytime and eight hours at night. Overall, entrepreneurs still insisted that the labor of children in industries was "an absolute necessity."[49]

Industrialists' almost united opposition to state intervention in general and to the proposed labor protection law illustrates their striking ability to join together in order to protect their entrepreneurial group interests. Evidence clearly suggests that during the debates, the industrialists who resisted the law's enactment developed certain rhetorical tactics that appealed to humanitarian notions. Their rhetoric emphasized not narrow entrepreneurial motivations but concern for the nation's well-being as a whole, specifically its economic and social interests. Contrary to its portrayals in some histories as "incapable" and "powerless" before the state,[50] the Manufacturing Council was very capable of defending the interests of its members and influencing the process of state decision making. Although the council technically remained under the authority of the Ministry of Finance (the council's chair was appointed by the finance minister), it formulated its policies quite independently from state and ministerial authorities. In this regard, the council provided a Habermasian "public sphere" for Russian entrepreneurs where they discussed various issues that affected their interests, formulated opinions, and promoted policies.[51] The council thus became an important mediator between the state and the Russian entrepreneurial community. Although the entrepreneurial community's successful opposition to restrictions on child labor in the face of predominant state support may be regrettable, the participation of entrepreneurs in the development and maturing of a civil society in late imperial Russia must be recognized.

Although the Finance Ministry's commission's initiatives were debated over a period of ten years, they remained dead letters. Even so, as mentioned, some of the commission's provisions served as the basis for the new regulations that restricted child labor in the mining industry and introduced regulations regarding free medical services and compensation for all workers. Moreover, the key provisions of these initiatives, as well as the debates about them, formed the criteria for later more successful efforts at factory legislation reform.[52]

Later Legislative Proposals and Public Debates

The legislative efforts and debates about child labor continued throughout the 1870s. During the 1870s, however, they took a new turn. Unlike the debates of the 1860s, which were confined mostly to only two groups, industrialists and state officials, the discussion during the 1870s involved a broader range of social groups. Legal and local government reforms of the 1860s, industrial growth, and the emergence and spread of new ideologies all played a role in bringing about these new developments. The newly introduced local representative governments (rural *zemstvos* and urban *dumas*) quickly involved themselves in the discussion of child labor. Contemporary periodicals also played a role in injecting the child labor issue (and related legislative projects) into the public arena. The late 1860s and 1870s also witnessed a sharp growth in workers' protest in the form of strikes and labor strife. A strike at the Nevskii Cotton-Spinning Mill in St. Petersburg in May 1870 ultimately involved eight hundred workers, making it one of the largest strikes of the era.[53] During the 1870s, strikes hit St. Petersburg, Moscow, Nikolaev, Riga, Odessa, and other industrial centers of the empire. Various contemporary political and economic theories penetrated Russia and stimulated the development of the workers' movement.[54] All these factors influenced the discussion of child labor and child labor protection legislation.

Concerned with the growing number of labor conflicts and the emergence of a labor movement, the minister of the interior reported in 1870 to Emperor Alexander II about the "urgent need" for a renewed legislative effort aimed at creating a comprehensive labor law.[55] Similar calls came from some provincial governors.[56] The emperor supported these initiatives. In October 1870, the imperial government organized a new commission to review the workers' and domestic servants' employment acts and appointed the State Council member Count P. N. Ignat'ev to head it. The appointment of Ignat'ev, a prominent statesman who from February 1872 would chair the Imperial Committee of Ministers, signified the high priority the imperial government assigned to the question of labor laws.[57] These events represent a new stage in Russia's approach to the labor question.

Meanwhile, labor legislation debates continued within industrialists' associations. Labor issues inspired lively discussions at the First Congress of Industrialists, which met in June of 1870 in St. Petersburg, several months before the appointment of the Ignat'ev commission. The Ministry of Finance specifically questioned industrialists about their attitudes toward labor legislation, evidently with the goal of having the Ignat'ev commission accommodate these views in its new proposal. The industrialists' congress held six sessions, the last of which centered on labor legislation and was open to the broader public. According to commentators, this session was attended by only four or five entrepreneurs; most of the entrepreneurs who had participated in the other sessions ignored this one. Regardless, many public activists, medical and educational professionals, and other reform-minded individuals who were interested in labor issues attended and actively engaged in the session's debates. The discussions focused on various labor-related issues, including the so-called workers' question. Although children's employment and limits on their work hours and nighttime work remained the most lively and controversial of all the issues raised at the session, participants also addressed other worker-related questions, including work, welfare, education, and morals.[58]

On the subjects of the minimum age for children's employment, work hours, and nighttime work, this congress shed little new light. As during the previous decade, opinion on children's workday and their minimum age for employment was sharply divided between supporters and opponents of existing legislative proposals. Some delegates suggested a total ban on industrial employment for children under fourteen years and educational opportunities and suitable work for juveniles between ages fourteen and sixteen. Other enlightened individuals wanted to prohibit employment for all juveniles in "perilous" industries, including rubber and tobacco. Such views came mostly from members of the reform-minded intelligentsia and representatives of the ruling elite. Several employers who represented technologically advanced factories that used steam engines and who probably believed that the law could bring them certain economic advantages also supported some restrictions.[59] Most business owners, however, maintained staunch opposition to a minimum age and a maximum workday. Their stated motivation was that they "still needed numerous auxiliary workers." Some opponents of labor laws felt that expanding industries experienced constant labor-force shortages, as a consequence of which a ban on child labor would have a negative impact.[60] Many entrepreneurs appealed to laissez-faire ideas and stated that the regulation of child labor was "an attack on the freedom of industry." Industry, they insisted, should remain free from government "regulations, restrictions, and inspections."[61]

Nevertheless, the congress's deliberations illustrate a new shift in the labor-legislation debates. It is interesting to note that although employers largely absented themselves from the sixth session's labor-legislation discussions, they employed a new strategy: to better protect their interests, they sent specially chosen delegates to the forum. These were well-educated and knowledgeable professionals, such as officers from industrialists' associations, economists, lawyers, and so on, who were capable of representing entrepreneurs' views and speaking for them.[62] Thus, the debates occurred mainly between these entrepreneurial agents on the one hand and reform-minded professionals—educators, physicians, economists, and labor-movement activists—on the other.

In general, the debates at the congress centered around two questions: the implications of the proposed child labor law for the nation's economy and the proposed law's likely effect on the material well-being of children and their families. Would these laws work for the betterment of the national economy and society as a whole? Would these laws benefit children and their parents? It comes as no surprise that each group posed as the best advocate of the nation's economic and social interests. Both groups displayed a profound degree of awareness of contemporary political and economic theories and easily manipulated these to bolster their arguments. As during the previous decade, the opponents of the legislation still maintained that the law would have a negative impact on the economy and society by hampering industrial development; increasing poverty among the lower classes; and, finally, placing Russian industry at an obvious disadvantage to foreign competitors, all of which utilized child labor to one degree or another.

The legislation's supporters countered these views by arguing that "the material benefit from the use of child labor [was] problematic" for working families because it reduced workers' monthly wages to minimal rates. They also argued that the alleged benefits for industries were illusory. For example, Doctor F. P. Vreden, a young political economist who had recently defended his doctoral dissertation, insisted that "the law must ban the employment of children under twelve, limit the workday for children at ages twelve through seventeen, and allow this employment only in industries not harmful to children's health."[63] Evidently well informed about modern economic theory, Vreden believed that the use of children in industries led to the reduction of wage rates for adult workers[64] and that, in the presence of child labor, "the working class [received] extraordinarily low wages insufficient to sustain families." Vreden advocated a "family wage" for adult workers, by which he meant a wage sufficient to support a family, eliminating the need for child labor. Appealing to humanitarian values, he maintained that child labor violated basic human rights and insisted that it must be banned, whereas the labor of women and juveniles, although permissible, should be strictly regulated by law.[65] One participant in the congress and bitter critic of child labor was Dmitrii Nikiforovich Kaigorodov, an early labor-movement activist and populist who later would join the first Marxist organization in Russia.[66] Kaigorodov sided with those voices that emphasized the need for prohibiting the labor of children twelve years of age and under. He also believed that the use of child labor in perilous industries was highly objectionable, to the point of being immoral.[67] Vreden's and Kaigorodov's positions received no response from child labor advocates. Those who opposed child labor laws and favored more or less unregulated child labor in the industrial economy remained firm in their views.

Another issue raised at the council that became controversial and ultimately escaped consensus was education for working children. (Even so, the very fact that the issue arose in a substantive way reflected a broadening of the discussion of child labor and childhood in general.) Although most discussants supported the idea of factory schools in general, they could not agree about funding for these schools. Some delegates suggested that employers must support factory schools, whereas most industrialists were not willing to take responsibility for financing children's education. Their deputies argued that education in factory schools must be paid for by some other means, since employers were burdened by other expenses. Some representatives of the business community suggested that small withholdings from workers' wages should finance factory schools.[68] The representatives of the reform-minded intelligentsia sharply challenged this proposal. For instance, in his response, Vreden commented bitterly that if industrialists employed children in a way "that brings them significant profits," they were obliged to spend some money on the children's welfare and schooling.[69] The supporters of labor laws connected the education issue to work-hour reductions in order to allow children the free time to attend school. One discussant maintained that without such reductions, the very idea of education would be useless "because after working fourteen or fifteen hours a day children would hardly find it possible to attend school."[70]

As reflected in the congress's debates, participants were concerned about measures for facilitating the intellectual and moral development of workers and creating an ideal type of worker. The debates illustrate, however, that there were no definite criteria for what a perfect worker should be. Some discussants, mostly the representatives of the ruling and business elites, emphasized improving the morals of workers and educating them by promoting Christian morality and religious values. For example, regarding the curriculum for factory schools, one delegate suggested that religious instruction and Christian morals would provide "the necessary basis for a disciplined worker." Ironically, some delegates even maintained that workers' "good morals" and "proper behavior" were facilitated by the long workday and child labor. They argued that restrictions on child labor would demoralize working families. One representative of the entrepreneurial community hypothetically asked, "What would working families do if children did not work until they are seventeen? What would women do? It is clear what they would do. These families would fall into drunkenness and poverty. There is no reason for banning children from work."[71] To support this view, another discussant noted that limiting work hours would lead to the reduction of workers' wages and, consequently, to material and moral deprivation. He stated that "the moral improvement [of workers] depends on their material well-being.... We should not put limits on but rather increase as much as possible all the means for raising wages." By "means" he apparently meant lengthy working hours.[72] Obviously, the ruling and business elites wanted to cultivate loyalty and obedience among workers in order to fulfill their vision of the perfect worker. They maintained that there should be "a close link between the intellectual and moral development of the worker and the interests of entrepreneurs."[73]

In contrast, progressive delegates at the forum—members of the reform-minded intelligentsia and the early workers' movement—perceived an ideal worker as an educated, aware, and socially active citizen. They were concerned about broadening workers' culture and suggested quite different conceptions of workers' morality.[74] Kaigorodov, for instance, noted that the moral health of workers actually lay in "the improvement of [their] material and physical well-being." This was the necessary basis for workers' culture. He pointed out that rather than teaching theology, factory schools should educate young workers in natural sciences, factory legislation, hygiene, history, and so on.[75] Kaigorodov suggested the founding of trade schools where working children could receive an education in general subjects, including math, geometry, and the Russian language, as well as industrial disciplines, such as drawing and industrial law. Factory schools, he thought, could also offer courses in church liturgy and gymnastics. He proposed that the state and entrepreneurs should finance these schools. Kaigorodov argued that it was the moral responsibility of the government and society to provide an education for working children. E. N. Andreev of the Russian Technical Society, a staunch advocate of limited working hours for children, argued that technical education could be effective only with broad education in general subjects.[76] As Kaigorodov's and Andreev's statements illustrate, progressive-minded middle-class reformers in Russia believed that the state should play a greater role in promoting the nation's welfare.

One author maintained in 1872 that "questions of the health, the material well-being, and the education of workers should be a prerogative of the government, whereas professionals, on the basis of data provided by research, should suggest the ways and means for best solving these questions, which are of great importance for the state."[77]

As mentioned, during the 1870s, various political ideologies began to penetrate Russia and influence the workers' movement. Entrepreneurs as well as state officials were concerned about the growing connections between workers and what they termed "undesirable" individuals and ideas. In 1870, Moscow manufacturers suggested founding a "Society for Workers' Welfare," which would concern itself with factory schools, libraries, and theaters for workers, as well as organize and maintain funds for worker' mutual financial assistance. The congress also suggested creating workers' credit associations and consumers' associations. By taking responsibility for maintaining workers' mutual assistance associations, entrepreneurs tried, in the words of one employer, to "prevent workers from attempting to organize themselves." Entrepreneurs clearly desired to exercise more control over workers' self-organization. Initially, these ideas attracted support from some officials at the Ministry of the Interior. The chief of the Third Department of the Imperial Chancellery, which concerned itself with policing society, worried, however, that the associations would be penetrated by "the currently multiplying followers of Flerovskii, Shchapov and Lassalle, who would use them in their own interests and thereby create a gap between labor and capital."[78]

Overall, the congress's debates did not come close to achieving a settlement between supporters and opponents of the legislation, especially as regards proposed limits on work age and work hours. In the resolution it sent to the Finance Ministry, the congress rejected the provisions on the minimum employment age and on maximum work hours for children. The congress found the idea of schools for working children "useful and desirable," but the entrepreneurial community did not want to take responsibility for funding the schools. The resolution also conveyed the entrepreneurs' desire that child labor laws correspond to existing norms in other countries. Obviously concerned about foreign competition, the Russian entrepreneurial community did not want any enacted law to place Russian industry at a disadvantage.[79]

Despite continued opposition from the entrepreneurial community, the legislative struggle continued—with, it should be noted, an ever broader agenda. In October 1871, the Ignat'ev commission reviewed the proposals put forward by industrialists and local governments and came up with a new legislative draft for a "Code on Personal Employment of Workers and Servants." In general, the Ignat'ev commission's legislative initiatives retained most of the provisions about the minimum employment age, workday, and nighttime work for children suggested by the Finance Ministry's commission in 1862. It also suggested some new approaches to labor regulation. The proposal outlawed the employment of children under age twelve and limited the workday for children age twelve to fourteen to eight hours. Children of that age could work four and a half hours at night per every twenty-four hours, whereas children between the ages of fourteen and eighteen could work ten hours during the daytime or eight hours at night. The draft would have obliged employers to provide employed children with education and medical care. Unlike the previous propositions, this draft contained more specific provisions for the implementation of the law and administrative penalties for employers who transgressed it.[80] (In Russia, as elsewhere, only strict oversight and real penalties would ensure the observance of labor regulations, even after such regulations entered the law codes. The provisions for implementation, entirely absent in earlier discussions, suggest a new seriousness on the part of legislators about the proposed regulations.)

As noted, during the late 1860s, Russia witnessed some labor unrest, as a consequence of which most local governors and state officials—that is, those who had to deal with these episodes at first hand—became ever more inclined to support labor protection regulations. For instance, in his 1871 report to Alexander II, the Moscow governor emphasized the need for new labor laws and for state factory inspectors who would supervise their implementation. The governor also supported the Ignat'ev commission's legislative effort and evinced some skepticism toward the industrialists' alleged concern for children's families. "Employers hardly ever acknowledge their exploitation of children," he insisted. "They shift the blame onto the parents as though they [the parents] force children to support their families." He believed that employers would on their own never exercise appropriate care for working children's welfare and that the government should be more involved in the matter of children's well-being.[81] Some government officials began to support even more decisive measures to restrict children's employment. When the legislative draft was discussed and reviewed in the Ministry of the Interior in 1872, the ministry suggested lowering the workday to six hours during the daytime and three hours at night for children age twelve to fourteen. For children between fourteen and seventeen years of age, the workday was to be limited to eight hours during the daytime and four hours at night.[82]

As in previous cases, most industrialists did not support the draft's provisions. When the business community learned about the Interior Ministry's changes to the draft, its associations immediately began to protest. The Moscow Stock Exchange Committee, an influential industrialists' association, called a meeting that produced a resolution that stated that the provisions regulating child labor would lead to "the inevitable elimination of all nighttime work, significant new expenditures for factory reorganization, and the rise of wages for adult workers because of the elimination of children from production. Replacement of children with adult workers would lead to the increase of production expenses, which will serve the interests of foreign competitors."[83] Furthermore, the chair of the Stock Exchange Committee, N. A. Naidenov, maintained that the proposed law "would be the kiss of death for the nation's industries . . ., whereas [at present] children perform easy tasks that cannot harm [their] health in any way."[84] Some industrialists also wanted to avoid taking responsibility for providing free medical services for their workers. Entrepreneurs were quite imaginative in developing arguments and excuses for their reluctance to support reform legislation. On the issue of children's health, one entrepreneur noted that "poor health among working children is caused by the extremely bad sanitary conditions of their home environment rather than by factory work itself."

"Complaints about exhausting child labor and its exploitation by the employers are groundless because," claimed one industrialist with obvious hypocrisy, "humane treatment of the weak is a characteristic of the Russian people."[85]

Of note is the fact that the debates about labor legislation during the 1870s received more publicity than those of previous years. This publicity and discussion were products of rapidly developing institutions and agencies of a civil society and its concomitant civic consciousness. Recent scholarship on state-society relations makes a strong case in general about the rise of civil society in prerevolutionary Russia. This phenomenon also found its reflection in child labor–legislation debates.[86] Newspapers and journals began to publish regular articles and essays about factory children, their working and living conditions, and the impact of factory labor on their health. The proposed regulations attracted attention and were discussed in the newly elected local representative bodies, the rural *zemstvos* and city *dumas*. Viewing the issue of children's employment differently from many industrialists, local governments mostly supported reformist ideas. Local governments especially approved limiting working children's employment age and working hours and providing them with education.

In 1873, the city *duma* of Ivanovo-Voznesensk (Vladimir province), one of the largest textile centers of Central Russia and the empire, suggested the introduction of a tax on local businesses in order to finance technical schools.[87] The *duma*, dominated by progressive-minded individuals of middling social levels, including entrepreneurs, favored technical education for factory children. In 1874, the Vladimir province *zemstvo* suggested very specific ideas about factory schools. It proposed that in factories where the number of workers reached one hundred, employers should establish schools for all working-class children (not just employed children). In the matter of the minimum age for starting employment and the workday for children, the Vladimir province *zemstvo* proposed that children under age fourteen should be banned entirely from employment and that the workday for children between fourteen and seventeen be limited to eight hours, with a required two-hour break for rest. The governor of Vladimir province also approved these progressive suggestions.[88]

Naturally, factory workers were among those who strongly advocated restrictions on child labor. For example, during the famous Krenholm Cotton Mill strike in 1872, workers demanded, among other things, limits on children's workdays and schools for factory children. Child labor was an issue in many other strikes as well.[89] Obviously, the use of children's low-paid labor reduced wage rates for adult workers, a factor that made child labor a matter of direct concern for them.

The ongoing public discussion during the 1870s began to create a more receptive climate for labor protection laws, even among some members of the entrepreneurial community. When the issue of child labor arose in 1874 in the Commission for Technical Education of the Imperial Russian Technical Society, this commission displayed considerable sympathy for labor protection and welfare laws. The commission, which seems to have been dominated by reform-minded individuals, included professors of economics, medical doctors, inspectors of technical schools, and a few entrepreneurs. Its head was the professor of economics and statistics Iu. E. Ianson. As noted earlier (see chapter 2), this commission gathered important comprehensive data about children employed in factories and about the conditions of their employment across the Russian Empire. The commission produced a thorough study of the impact of factory labor on children and concluded that the health of most children employed in factories was poor. It worked out specific legislative recommendations for imperial lawmakers. Three members of the commission suggested eleven years as the minimum age for beginning factory employment, whereas the other twelve members advocated twelve years as an absolute minimum.[90]

The commission emphasized the moral and medical aspect of the use of children in industries. In its resolution, with full references to contemporary medical research, the commission attempted to provide a detailed explanation of why the employment age should be limited to twelve and the working day to eight hours. Physicians who participated in the commission maintained that the physiology of children under age twelve was "so weak that any continuous work is very harmful. At this age, children cannot pay enough attention and exercise necessary caution [while working with machinery] and therefore are easily vulnerable to the various dangers this machinery may pose,"[91] an observation that finds support in recent research on child development (see chapter 2). In essence, the resolution implied that the industrial employment of young children, persons who had not attained the necessary physical and mental maturity, was immoral and should be prohibited outright.

On the issue of education for juvenile workers, the Commission for Technical Education came up with concrete and quite progressive ideas. It suggested that factory schools should be set up no more than four kilometers (about two miles) apart in all locales where the number of factory and shop workers approached five hundred people. Additionally, the commission proposed to introduce a tax on all businesses at the rate of 0.5 to 2.0 percent of the amount spent on workers' annual wages in order to organize these schools and a tax on all workers at the rate of 1.0 percent of their salary in order to provide free education. The commission also suggested that the employment of children between the ages of twelve and fifteen should be utilized only if they attended school at least three hours a day. The commission specified that employers should not require juveniles between ages fifteen and seventeen to attend school, nor should they prevent such children from attending school. The commission proposed requiring that workers between ages fifteen and seventeen attend factory schools if they did not have at least two years of previous public schooling.[92]

Similar ideas on the schooling of working children received considerable support at the Council of Machine-Making Industrialists, which met in 1875. This council, as opposed to other entrepreneurial associations, welcomed the enactment of child labor laws.[93] During the late 1870s, the Society for the Support of Russian Industry and Commerce also considered the recommendations of the Commission for Technical Education. The society agreed about most provisions on work safety and working children's education but suggested ten years as a minimum employment age, pointing out the British and French examples. In most

European states where child labor laws existed, the minimum age for employment was usually set at ten or twelve and the workday for children under age fourteen limited to ten hours.[94]

In 1874, Ignat'ev's proposals, along with public opinion on the matter, were reviewed by a specially appointed committee that included the representatives of various ministries, members of the nobility, representatives of provincial and local government, and representatives of six large enterprises, with the minister for state possessions, Count P. A. Valuev, as the chair.[95] Some government officials believed that Ignat'ev's draft attempted to address too many aspects of labor at once and that, in order to expedite their introduction, the draft's provisions should be divided and then gradually enacted according to appropriate priorities. Thus, the Valuev committee retained all the provisions it believed to be of the highest importance and that needed to be enacted first. It excluded from Ignat'ev's draft sections on workers' associations and labor-arbitration courts, which, as noted, caused some tension within the Ministry of the Interior.

Among the various issues, the committee gave child labor the highest priority. It suggested limiting the maximum workday for children between ages twelve and fourteen to six hours a day and three hours at night, and for juveniles between fourteen and seventeen to eight hours a day and four hours at night.[96] As previously mentioned, this reduction of work hours for children had already been suggested by the Ministry of the Interior. Regarding the minimum age for employment, the committee suggested twelve years as the appropriate age to begin working in factories and ten years to begin an apprenticeship.[97] Employers could not require employed children to do work that did not fit their age and strength. The provisions on schooling obliged employers to "provide employed children with the time for attending schools." The committee suggested penalties for violations from fifty kopecks to ten rubles, depending on the given violation.[98]

Nine members of this committee, mostly representatives of the business community, submitted a "special opinion" about child labor regulations. They agreed with the minimum employment age but suggested increasing the maximum workday for children between ages twelve and sixteen to nine hours, arguing that the six-hour workday limit for children was "impractical and unrealizable." They maintained that in those countries where child labor was regulated, minimum workday provisions were usually unworkable and inoperative.[99] The representative of the Finance Ministry supported their opinion.[100]

Again in 1875, as in 1862, this new proposal was sent to various industrialist associations for review and discussion. The Valuev committee requested that local and provincial governments and various public organizations respond to questions about the new legislative proposition. Once again, the majority of entrepreneurs did not support the child labor provisions. When the proposed law was discussed at a specially appointed commission of the Riga Stock Committee in 1875, the commission suggested limiting the minimum employment age to ten years and the workday to six hours for children between ten and thirteen years of age. The commission, however, approved the idea of mandatory schooling for children under age thirteen. The Ivanovo-Voznesensk Committee for Trade and Industry similarly supported the idea of education for factory children but proposed limiting the minimum employment age to ten years. In 1881, the Society for the Support of Russian Industry and Commerce submitted a statement that also suggested ten years as the minimum age for employment. This issue aside, the society revealed positive and progressive attitudes about factory schools for children.[101] Nevertheless, by the late 1870s, broad public opinion—including most state officials, members of local representative government, the reform-minded middle class, and most members of the intelligentsia—had long anticipated the end of child labor, a practice they considered morally unacceptable and downright evil. Unregulated child labor in Russia was finally approaching its end.

Although the entrepreneurial community still mostly rejected crucial provisions regarding child labor, the 1870s did witness a significant transformation of public attitudes about the issue. The involvement of reform-minded individuals, members of the workers' movement, and economic theorists, such as Kaigorodov and Vreden, in the discussions epitomized the growing public concern about child labor and labor protection legislation. Although during the 1860s, some educators and medical doctors had taken part in labor-law discussions, industrialists and state officials had predominated. In contrast, during the 1870s, child labor and labor legislation became broader public issues. Increased publicity about children's industrial employment and the impact it had on children increasingly outraged public opinion in Russia. This transformation of attitudes about child labor is perhaps best reflected in two starkly contrasting statements made by the Vladimir provincial authorities. In the early 1860s, the governor of the province had expressed absolute support for children's employment in factories and insisted that "children's immaturity" was not a sufficient cause for restricting child labor. In 1878, the new governor of the same province wrote that "one of the evils that marks industrial areas is the use of children age ten and under [for work]."[102] This contrast signifies the fading away of the old perception of child labor as a means of apprenticeship in favor of an entirely new concept of childhood and education. The idea of child labor as a form of education was yielding to the idea of sending children to schools for an education.

During the 1870s, the child labor issue, and discussion thereof, encouraged the emergence of a broader social-welfare reform movement. Many concerned and reform-minded contemporaries expressed their opinions in journals and newspapers. Numerous articles in contemporary periodicals addressed the issues of factory labor, labor protection laws, and workers' welfare in general. For example, in an 1871 article in the popular medical journal *Arkhiv*, one author (who wrote under the pseudonym "P.") called for the introduction of labor protection and sanitary laws and for the creation of a system of independent factory inspectors and physicians. He maintained that "we still have quite insufficient organization of sanitary control over factories, plants, workshops, and so on, because of the absence of laws that would adequately protect the life and health of workers, as well as because of the absence of personnel responsible for control of sanitary conditions of workers in industry."[103]

According to the historian of medicine A. P. Zhuk, this author was probably S. P. Lovtsov, a medical doctor and public activist. In this and other articles, he laid out a whole program of responsibilities for factory medical inspectors. These responsibilities centered on control over employment and labor, including for women and children, and the supervision of education for employed children.[104] In an 1872 article in *Znanie*, Lovtsov, writing under his own name, emphasized that "a more radical means for protecting workers' health would be the rise of wages and the decrease of working hours.... This would reduce workers' time in workshops and thus cut sickness and mortality rates among them." He also supported the minimum employment age and argued that the law should ban children and juveniles under age eighteen from employment in industries and from certain kinds of work that could jeopardize children's health.[105] Lovtsov and other progressive-minded individuals initiated a concerted attack on previously predominant views among entrepreneurs to the effect that long working hours and child labor were the best means of raising incomes and promoting the well-being of workers' families.

From the 1870s on, various periodicals began to publish regular articles on the working and living conditions of working children. These publications exposed to public view child labor and conditions among children in industry.[106] In all likelihood, educated society had previously been completely unaware of these matters. Many publications now devoted whole issues to child rearing and children's education. For example, in its section "The Domestic Observer," the political and social journal *Vestnik Evropy* regularly published articles about conditions among children in industries.[107] The eminent educator V. I. Liadov published his famous manual on child rearing and upbringing.[108] The medical journal *Arkhiv* devoted many pages to childhood and children's health. Most of these articles portrayed child labor as an evil practice that must be outlawed. Many doctors devoted research to and published studies on the issues of children's diseases and mortality. For instance, V. S. Snegirev defended a doctoral dissertation entitled "About Mortality among Children under the Age of One" with the Medical Surgical Academy. In his polemic against some authors who emphasized race and climate as determining factors in children's deaths, Snegirev concluded that child mortality primarily reflected the social conditions endured by the mass of the population. He emphasized the importance of education and better material conditions as the key factors in a population's well-being.[109] This public discussion of the broad issue of childhood produced an environment that favored the introduction of child labor laws.

Another factor in the transformation of public opinion was popular literature, a particularly potent force in Russia. During the late nineteenth century, children and childhood occupied a special place in Russian literature. Many authors exposed and, in effect, denounced abuses against children employed in factories, workshops, and domestic service. In his 1888 short story "Spat' khochetsia" ("I Want to Sleep"), Anton Chekhov describes a thirteen-year-old girl, a babysitter and maid in a craftsman family. An unbearable longing for a little sleep becomes an obsession for this overworked and exhausted child. Finally, either in her fragile dream or in some bleary reality, she realizes that the "force that binds her arms and legs, that chains her life" and prevents her sleep, is the child. Chekhov continues: "A mistaken thought" seizes her—"kill the baby and then sleep, sleep, and sleep." And the girl strangles the child.[110] In another story, "Van'ka" (1886), Chekhov recounts the history of a nine-year-old boy, Vania Zhukov, who has been sent to an apprenticeship. In a letter to his grandfather, addressed briefly "to the village, to grandfather, Konstantin Makarych," the boy complains about the severe abuses that he has to endure from his master. As he remembers his village life, the boy begs his grandfather to take him back home to the village.[111] Unfortunately, as the address may suggest, this letter never reaches its destination. Perhaps these children's total hopelessness, echoed in Chekhov's and other writers' stories, is artistic exaggeration. Nevertheless, the stories represent growing concern among the educated public about working children. Late nineteenth-century writers perceived child labor as a wicked practice. Like other factors mentioned earlier, literature's condemnation of child labor signified new perceptions of childhood.

This growing public interest in children and childhood also influenced the development of a specialized literature for children. Although the origins of literature for children in Russia dates back to the late fifteenth century, during the late nineteenth century, children's literature became a prominent Russian genre.[112] Several children's series, including "Children's Books for Sundays," D. F. Samarin's "Library for Children and Youth," and A. S. Suvorin's "Low-Priced Library," among many others, emerged as popular periodicals affordable to children of the lower social strata. Great writers such as Leo Tolstoy, Fyodor Dostoevsky, Leonid Andreev, Maxim Gorky, and V. G. Korolenko wrote stories and novels for children. Many authors devoted single volumes especially to young audiences.[113] Children's literature tried to encourage children's curiosity about the world and cultivate a love for reading and learning. It emphasized school education as a primary priority of childhood.

In summary, the debates about labor laws created an atmosphere that favored the introduction of the labor protection and social-welfare laws that in reality marked the late imperial decades. In addition, these ongoing debates illustrate two important yet underreported aspects of imperial Russian development. First, the debates reveal the remarkable transformation of the Russian business community during the second half of the nineteenth century into a vigorous and powerful social entity capable of influencing state policies. The involvement of the business community's associations in a two-way dialogue with the autocratic government delineates new boundaries between state and society. Second, concomitantly, the debates display the process of lawmaking in imperial Russia in a quite different light than is usual. Rather than being a product of one or another top-level bureaucrat or committee, laws arose from broad public discussion and compromise among various social groups that in this, as in other cases, resulted in legislative efforts or in cancellation or alteration of such efforts.

Although most legislative propositions about child labor did not become law during the 1860s and 1870s,[114] they, along with contemporaneous public debates about child labor, laid important intellectual and juridical foundations for the 1880s laws regarding children's employment, work, education, and welfare. Ultimately, the proposals and the debates about them facilitated the introduction of actual laws in the decades to come. These laws, their implementation, and their significance will be discussed in the following chapter.

4 ⚙ Factory Children
Politics, Education, and the State

THE LONG PUBLIC DISCUSSION of the 1860s and 1870s about child labor in industry finally yielded the 1882 law, the first decisive act to restrict the industrial employment of children. The following years and decades witnessed the introduction of labor protection and welfare legislation concerning all industrial workers. Starting with the 1882 law, the government limited the employment of children in all private industries. The laws banned the nighttime labor of children and their labor in perilous industries, including underground work in mines. Simultaneously, the state introduced mandatory schooling for children hired for factory work. Again, it is worth emphasizing that, beginning with child labor protection laws, labor legislation expanded its scope to include other categories of workers and other labor-related issues. During the late imperial decades, a series of laws limited the workday, legalized strikes and workers' unions, introduced health care and state-sponsored medical insurance for all workers, and established pensions for some categories of disabled and retired workers. In order to implement labor protection and welfare laws, the state instituted the factory inspectorate. All these laws directly applied to hundreds of thousands of children employed in industry.

What did these laws accomplish? What happened to those children who were banned from employment and to those allowed to take factory jobs?

The 1882 Child Labor Law and Its Implementation

In December 1881, the minister of finances, N. Kh. Bunge, known as a liberal official, forwarded the new legislative draft "On the Labor of Children and Minors" to the Imperial State Council for approval. After revisions in various legal departments of the State Council, in June 1882, the council and the emperor finally accepted and approved the draft. In legal and historical literature, it became known as the June 1882 law. (The main points of this law are included in the appendix.) The law barred children under age twelve from employment in "factories, plants, and manufacturing establishments." It limited work for juveniles between ages twelve and fifteen to eight hours a day, excluding time for breakfast, lunch, dinner, school, and rest. Work could not last more than four consecutive hours. It also prohibited work between nine P.M. and five A.M. in the summer and spring, and between nine P.M. and six A.M. in the fall and winter, as well as on Sundays and important imperial holidays. The law also banned the employment of children of the specified ages in "industries harmful to children's health." The ministries of Finance and the Interior were to issue a list of such industries, which they provided by June 1884. The provisions of the law obliged employers to provide their teenage workers at least three free hours a day, or eighteen hours a week, in order to attend public schools or their equivalent.[1] In order to provide businesses with time to accommodate the law's provisions, the government scheduled the enactment of all statutes that concerned children's employment for May 1, 1883.[2] Thus, after almost two decades of public discussion, the state finally imposed universal restrictions on child factory labor.

The 1882 law, as well as later laws that applied only to certain kinds of businesses, distinguished three age categories of children. These categories included children under twelve years of age, who were banned from employment; children between the ages of twelve and fifteen (defined as *maloletki*); and juveniles from fifteen to sixteen (*podrostki*). Children in the latter two age categories, of course, were suitable for employment. Individuals age seventeen and above were considered adults. Child labor protection laws introduced after 1882 applied primarily to children between ages twelve and fifteen and to a lesser extent to juveniles age fifteen or sixteen.[3] The 1882 law concerned factory labor and also extended its reach to all private businesses equipped with steam colanders, steam or mechanical engines, and machines and lathes, and to all establishments that employed over sixteen workers.[4]

In all enterprises that fell under the 1882 law's scope, it provided for a system of state control over working conditions for children. By June 1884, the government had organized the fifty-eight provinces of European Russia into nine "industrial districts." In each district, an office of factory inspectors supervised the implementation of laws "that regulated employment, work, and education of juvenile workers and examined, with the aid of members of the local police offices, transgressions of this legislation." The government created the Moscow, St. Petersburg, and Vladimir industrial districts in 1882 and during 1884 added the Voronezh, Kazan', Kiev, Kharkov, Vilna, and Warsaw districts.[5] The provisions for factory inspectors, however, did not apply to state-owned industries or to privately owned mines, since these establishments had their own inspectorate systems and regulations. Control over the implementation of labor laws in these latter businesses belonged to their administrations or, in the case of mines, to the Mining Administration.[6] In addition, the Asian portion of the Russian Empire, especially western and eastern Siberia, which had a significant number of mining and metallurgical industries that employed children, also remained outside of the factory inspectorate's jurisdiction. Most Siberian mining also had its own inspection system, of mixed effectiveness, introduced during earlier decades.[7]

Each of the industrial districts consisted of a number of imperial provinces of European Russia and initially had one inspector and one assistant, obviously quite an inadequate number.[8] Factory inspectors were subordinate to the Ministry of Finance. Annual salaries for inspectors were 3,000 rubles, and their assistants received 1,200. According to an 1882 editorial in *Vestnik Evropy*, these were "meaningful" sums. In order to maintain the inspectorate, the state made an annual appropriation of 78,500 rubles and introduced a tax on industries, which ranged from 5 to 100 rubles depending on the number of people employed in the given firm.[9] Factory inspectors were qualified persons of the economic, legal, medical, and engineering professions—for example, the doctors of medicine P. A. Peskov and F. F. Erisman and the prominent economist I. I. Ianzhul, a professor at Moscow State University. The proficiency of factory inspectors was unstained. For instance, Erisman was a professor of medicine at Moscow State University and the founder of the first medico-epidemiological station in Moscow in 1891. Further, the academic, research, or professional activities of many of these appointees related to factory labor or to workers. Some of them had actually served as provincial supervisors of sanitary conditions in factories and were therefore quite familiar with industry. The chair of the Committee for Technical Education of the Russian Technical Society, E. N. Andreev, who had taken an active part in the preparation of the 1882 law, became the first chief factory inspector.[10] Later, some contemporaries noted early factory inspectors' "high qualifications and professionalism."[11]

In late 1882, the Finance Ministry notified employers about the new law through the auspices of local police offices. The ministry sent out circular letters to private businesses informing them of the introduction of new factory labor regulations and of the factory inspectorate. The letters required employers to sign and return a memo confirming that they had received and read the information.[12] In March 1883, Chief Inspector Andreev asked employers to communicate to him their opinion about the newly introduced labor regulations. Most employers reported that they did not see any major obstacles to the law's enactment. Many pointed out, however, that the provisions regarding minimum employment age and nighttime work were troublesome and would require time to make certain adjustments. Employers also emphasized that the regulations must apply to all businesses across Russia simultaneously so that the law provided equal conditions for all owners.[13] As noted, by the time of the new laws' adoption, much of the opposition came from owners of smaller, traditionally organized businesses. Some industrialists complained that the immediate enactment of the law would place many families, which had already arrived at factories with their children, not to mention entrepreneurs themselves, in a "quite awkward situation."

In April 1883, Andreev reported to the finance minister about some employers' concerns that "after the Easter holidays many workers would return with their families from the countryside and may find themselves forced to support underage family members with their own means.... In addition, many businesses employ underage orphaned children who may find themselves without any means of subsistence."[14] Trying to accommodate industrialists' concerns, the finance minister, N. Kh. Bunge, wrote to the State Council that "because of the great significance that the labor of children has attained in some businesses," the immediate enactment of the law would create problems for employers.[15] He asked the council to delay the enactment of the children's employment provisions for one more year. Thus, the law's provisions, which were to be implemented on May 1, 1883, were actually brought into effect a year later, on May 1, 1884. The provisions that concerned the introduction of the factory inspectorate, however, went into force in June 1882.[16] The postponement of the enactment of the provisions on children's employment allowed businesses one final year to make necessary changes to production and labor organization as a basis for dismissing all children under age twelve.

In addition, as a result of industrialists' initial pressures, in 1884 the government introduced some provisional adjustments to the 1882 law. Employers still complained that some of the law's terms did not fit the labor and production processes normally practiced in their businesses. For example, as noted, the workday in most textile mills lasted twelve hours in two six-hour shifts. This type of organization conflicted with the new law's terms that limited the workday for children to eight hours and required children's work to last no more than four consecutive hours a shift. An 1884 provision of the child labor law allowed enterprises that operated in six-hour shifts to have children work six consecutive hours instead of four. The total workday for these children, however, was limited to six hours a day. In addition, in 1884 the government allowed some industries, including glass making, to employ ten-year-old children as apprentices and "assign them work appropriate to their strength." The definitions of "appropriate work" and children's "strength" were laid upon employers and factory inspectors. These provisions were introduced as temporary measures until May 1, 1886.[17]

Regardless of these problems, with the completion of the industrial districts by October 1884, the government filled all eighteen factory inspectors' and assistants' positions. Inspectors started their work in January 1885.[18] The chief factory inspector and district inspectors of Moscow, St. Petersburg, and Vladimir had been appointed earlier, in late 1882. Starting in 1882–83, these individuals began to collect data on their districts' industries and factory labor, an effort joined by the other inspectors during 1884–85. Most owners welcomed inspectors to their businesses. Some employers, however, met them with hostility, and a few did not even want to let inspectors enter their enterprises and contact workers. In the Kiev district, some cases of coarse treatment of inspectors by employers occurred. In St. Petersburg, a number of employers initially refused to admit inspectors. A few Moscow owners tried to dispute the inspectors' right to question their workers and, like their Petersburg counterparts, refused to admit inspectors to their businesses.[19] Some industrialists still believed that the introduction of factory inspectors was a crude state intervention in "the private affairs" of entrepreneurs and their businesses, a perhaps not unexpected assertion of laissez faire among Russia's nascent capitalists. The Voronezh district inspector remarked that some employers were posing questions about "what business this is of the government and why there is suddenly so much attention to children." These employers believed "that [child labor] was such a minor issue that it should not concern any respectable government." Entrepreneurs of this cast of mind resisted factory inspectors' visits.[20] These cases were, however, uncommon. Most employers, as mentioned, cooperated with inspectors.

In any event, during 1882–85, inspectors and their assistants assumed responsibility for over 25,913 businesses that fell under the technological and juridical scope established by the 1882 law. Overall, these businesses employed 870,969 workers and spread across an enormous territory of over four million square kilometers. Thus, on average, each inspector or his assistant was responsible for supervising about 1,440 enterprises, a daunting figure. The Finance Ministry provided factory inspectors and their assistants with travel funds, although inspectors claimed that these funds at first came with some delays. By the middle of 1885, the inspectors and their assistants actually inspected and gathered data from 4,897 enterprises, a considerable number, but only 20 percent of the total for which they were responsible.[21]

Although in late 1882 local police offices informed local businesses about the introduction of child labor regulations, when inspectors came, some employers claimed to be completely unaware of the new law's existence. Investigation revealed, however, that many of these claims were groundless: entrepreneurs in fact had received the information about the new laws and signed a confirmation memo. Clearly some employers claimed ignorance as an excuse for continuing their former practice of employing underage children.[22] In 1885, in order to avoid uncertainty and facilitate awareness among employers and workers, the Finance Ministry published a single edition of "Collection of Decrees about Underage Workers Employed in Factories, Plants, and other Manufacturing Establishments," at a price of twenty-five kopecks.[23] Factory inspectors believed that this price was "low enough" and affordable for most workers. In addition to all legislative acts regarding child labor that had appeared since the 1882 law, this publication included two separate leaflets, printed in larger letters, that contained "Rules for Employers" and "The List of Harmful Industries," as well as samples of data sheets for factory inspectors. Instructions required factory owners to post the leaflets in factories in open places accessible to all employed people. For regions with large non-Russian populations, copies of the volume were published in native regional languages, in addition to Russian. Official and popular periodicals also published information about the new labor regulations.[24] In addition, when they visited factories, inspectors informed their owners or managers about the new child labor regulations. They explained the meaning of these regulations and suggested what should be done in each specific case in order to implement them.[25]

In December 1884, after the first factory inspectors had gained some experience, the Finance Ministry, in coordination with the inspectors, worked out detailed instructions for factory inspectors and employers. The instructions tried to eliminate vagueness in interpretation of the law and to clarify and facilitate its implementation. Provisions of the instructions required owners to employ children only with documents that identified their age. Employers were responsible for keeping copies of these documents in factory offices and presenting them to factory inspectors on demand. The instructions required employers to maintain records about all employed individuals under the age of seventeen who were allowed for employment and report to the district inspectors about their further intentions to use these persons' labor. The instructions suggested what kinds of information inspectors should gather about businesses and workers and how to report this information. They contained sample tables that inspectors were required to fill out and return in their monthly reports on businesses that they had inspected during the month in question. In addition, the instructions required inspectors to control the implementation of the Medical, Fire Protection, and Building codes in industries, along with child labor laws.[26] In other words, the government demonstrated a clear intention to enforce the implementation of the new labor laws.

Some employers apparently attempted to evade the law by manipulating its language and finding rhetorical loopholes in its provisions. For example, when inspectors visited factories and saw children under the age of twelve on the shop floor, employers sometimes maintained that these children were not workers but had simply accompanied their fathers or relatives. Employers claimed that they could not stay at home because there was nobody to take care of them, a not entirely implausible assertion in light of the total absence of childcare institutions. To clarify the ambiguity, the instructions stated that "even the presence of children in a working room constitutes [a situation in which] they are performing work" and that such children must be considered workers. Thus, the presence of children under the specified age in places where work was conducted constituted a violation of the 1882 law.[27]

In order to reinforce the implementation of the 1882 law and later labor acts, in 1884 the government added the Penal Code, with additional provisions that specified and increased sanctions for violations of labor regulations. According to these provisions, employers who transgressed child labor laws could be sentenced to "no more than one month" of imprisonment or fined up to one hundred rubles. The same penalties applied to employers who failed to provide their employed children free time to attend schools.[28] Of course, one hundred rubles was quite a trivial sum for many entrepreneurs. In cases where the law was violated, factory inspectors could adjudicate the violations in coordination with the local authorities or simply file reports with local police or courts, which would then presumably follow up on the matter. The imposition of penalties, inadequate as they were, did signify to entrepreneurs the government's serious intent: Russian subjects, like people everywhere, habitually ignored or otherwise evaded laws that lacked specific focus and penalties.

As described earlier, the introduction of the 1882 law resulted from the debates of the 1860s and 1870s, which had created a favorable background for its implementation. In fact, educated, reform-minded society had long awaited the law and welcomed it when it finally arrived. In August 1882, the journal *Vestnik Evropy* noted that "the need for protecting children employed in factories has long been established not only by society and literature but by the government."[29]

Nevertheless, humanitarian considerations and increasing reformist attitudes did not alone carry enough weight to push through the needed labor protection legislation. In the end, it was the economic slump of the late 1870s and early 1880s caused by overproduction that facilitated the introduction of the 1882 law. Industrial output heavily exceeded market demand for goods. This caused businesses to reduce their production, which in turn created unemployment. Many businesses laid off a portion of their workers, including many children. Wages of industrial workers declined, partly as a result of the reduction of working hours. Large industrial centers witnessed a wave of workers' protest and strikes.[30] The factory inspector of the Moscow industrial district, Erisman, wrote that "had the introduction of the 1882 law not occurred during the industrial crisis of the early 1880s, the struggle of industrialists against the law would have been more energetic." Indeed, some employers even suggested the complete elimination of nighttime work as "the best measure" for overcoming the ongoing crisis.[31] Thus, during the early 1880s, unlike during the previous decades, despite a few individual cases of antagonism, no strong consolidated resistance to child labor regulations arose. This undoubtedly facilitated the laws' implementation and enforcement. The government's efforts to regulate labor relations and provide welfare for workers continued during the following decades. The introduction of the 1882 law signified the beginning of a coherent process of labor protection legislation in imperial Russia, a little-noted process that continued throughout the regime's last decades.

Later Imperial Russian Laws on Employment, Labor, and Welfare in Comparison to Those of Other Industrializing Countries of Europe

After the implementation of the 1882 child labor law, the government ultimately extended its concern toward all workers, adults as well as children, regardless of their age and gender. New legislative acts further restricted the labor of children between ages twelve and seventeen, scrutinized the implementation of labor laws, introduced compulsory education for employed children, and addressed the employment of women. Later legislation also established the maximum workday and instituted medical care, state-sponsored medical insurance, and disability compensation for all factory workers. A series of laws during 1905 and thereafter legitimized strikes and workers' associations. In order to enforce compliance with all these laws, the government dramatically increased the number of inspectors, as well as the scope of their authority.

In June 1884, the Finance Ministry issued a list of types of industry and work where the employment of children under age fifteen was prohibited. This was an extensive list of thirty-six industrial spheres with specified occupations and workshops where children could not be employed. The list included certain textile operations; oil refineries; mills that processed minerals; chemical plants that produced acids, paints, and varnishes; spirits distilleries; and slaughterhouses, among many others. In some businesses, children under age fifteen were allowed to perform only certain specified tasks. In bakeries, for instance, they could only pack and carry bread; contact with ovens and other machines was forbidden.[32]

On June 12, 1884, the government introduced a law regarding mandatory schooling for children between ages twelve and fifteen employed in industry who had not yet completed at least a one-year program of public schooling. The law required these children to attend schools and to complete a one-year curriculum at a public school or its equivalent. The law advised but did not oblige factory owners to open and maintain factory schools, if public schools were remote from factories and not available to working children. The law's statutes laid the responsibility for organizing factory schools on factory inspectors and local education authorities. The law obliged the Ministry for People's Education to develop a curriculum and teaching plans for factory schools.[33] Although the organization of factory schools was nonobligatory, the law nonetheless made employers responsible for children's education. Employers either had to maintain a factory school, if no public school was available nearby—true for many enterprises—or hire only those children who had already received the required education.

The laws of 1885 and 1886 prohibited nighttime work for children under age seventeen and for women in the cotton, linen, and wool industries and in mills that processed mixed fibers considered harmful to workers. Local and provincial authorities, however, retained the right to admit teenagers and women to nighttime work in some exceptional cases. With the agreement of the minister of the interior, the minister of finances reserved the right to extend this legislation to other industries.[34]

In 1886, the state introduced the first universal law "on factory employment and on relations between manufacturers and workers." This law included all of the abovementioned provisions regarding child labor and also broadly addressed adult industrial labor. The law regulated employment contracts and relations between workers and employers and extended the responsibilities of factory inspectors toward all industrial workers regardless of age. The latter provision, however, applied initially to only three of the most industrialized districts of imperial Russia—that is, Moscow, St. Petersburg, and Vladimir. During the 1890s, the government extended the law's scope to other industrial districts. The law increased the number of inspectors by adding ten new assistants' positions. The law also included provisions that obliged employers to provide workers with certain basic medical services.[35] Thus, almost all the measures regarding child labor proposed and discussed during the earlier decades became law during the 1880s. By the 1890s, many of these provisions applied to other age groups as well. The debates about child labor therefore formed the foundation for the enactment of child labor laws, a process that spanned several decades. This process served as a template for universal labor protection legislation.

By the mid-1880s, the economic crisis of the early 1880s began to recede, and industry began a recovery. The economic revival and the reopening or expansion of many businesses demanded a larger workforce. At this point, the labor laws and inspection system came under vigorous and consolidated attack from employers. For example, in 1887, the Moscow Association for the Support of Russian Industry complained to Finance Minister I. A. Vyshnegradskii that with the introduction of the factory inspectorate there had occurred many "disagreements and conflicts between inspectors and employers." Industrialists stated that "the law placed factories at the mercy of persons [inspectors] who do not know the industry and its needs."[36] Employers demanded the elimination of certain provisions regarding child labor. Individual owners sent letters to the government requesting temporary exemptions from the child labor laws. For instance, in 1889, the owner of the Murakov firm asked the Ministry of the Interior to grant his business a five-year moratorium on labor laws, stating that his recently established enterprise was "relatively small in production volume." The ministry, however, refused to grant the request.[37]

Nonetheless, under constant pressure from industrialists, the government agreed to introduce some relaxations of the existing law. In 1890, the government allowed children between ages twelve and fifteen to work on Sundays and important imperial holidays, with the agreement of factory inspectors. The government also increased the workday for children to six consecutive hours in businesses that utilized a twelve-hour workday in two six-hour shifts. (As mentioned earlier, a similar provision, introduced in 1884 as a temporary measure, was in force until May 1886.) In industries that operated eighteen hours a day in two nine-hour shifts, the workday for children was increased to nine hours. This was done in order to reconcile working hours for children with the workday of adult workers whom they assisted. Regardless, the concessions did not go so far as to eliminate the outright ban on the employment of children under twelve.[38]

Furthermore, despite increased opposition from employers during the late 1880s, the legislative effort to further restrict child labor continued throughout the 1890s and into the first decade and a half of the twentieth century. In 1892, the government introduced restrictions on the labor of children and women in the mining industry. This law banned children under age fifteen and women from nighttime work and from work inside mines and underground. The law specified that nighttime work was work that took place between nine P.M. and five A.M. in spring and summer and between nine P.M. and six A.M. in winter and fall. The workday in the mining industry for juveniles between ages fifteen and seventeen was limited to eight hours.[39] In 1897, the government introduced "The Statute on Rural Handicraft Workshops," which extended all existing labor regulations to rural handicraft enterprises. (One wonders about the enforceability of this worthwhile endeavor.) In the same year, another law limited the workday for adult workers to eleven and a half hours during the daytime and ten hours at night and to twelve hours in businesses with a continuous production cycle. Introduced at first in the nine industrial districts, all such factory labor regulations soon spread to most other territories and provinces of imperial Russia. The government also organized new industrial districts in Azerbaijan and Georgia.[40] The state specifically placed responsibility for implementing all these laws upon factory inspectors. An appropriate conclusion would be that the interaction between entrepreneurs and the state as regards child and other forms of labor was dynamic, interactive, and dialectical. The end result was constantly increasing state control over and limitation of labor practices, especially for children.

As noted, the labor regulations introduced after 1886 dramatically expanded the factory inspectorate and its area of responsibility. In 1886, the inspectorate consisted of twenty-nine individuals, including nine inspectors, nineteen assistants, and one chief inspector. The number of inspectors was obviously insufficient to provide for effective oversight of labor laws. In order to reinforce the factory inspection system, the government drastically increased the number of inspectors. By 1894, the factory inspectorate included 18 senior inspectors, 125 inspectors, and 20 assistants. The position of chief inspector was eliminated. The law of 1897 introduced 20 new positions for factory inspectors and 3 for factory *revisory* (supervisors whose functions mirrored the former chief inspector's), thus increasing the inspectorate to 185 persons.[41]

State legislative efforts to regulate labor relations and introduce labor protection continued during the early twentieth century. Despite the extensive legislation and statutes to enforce adherence during the 1880s and 1890s, many legal issues regarding factory labor relations remained unresolved. For example, such crucial questions as workers' associations and labor unions, not to mention workers' unemployment compensation and medical insurance, remained open. In 1905, under the grave pressure of massive labor unrest that year, the government created a commission to reform and extend labor legislation and appointed Finance Minister N. V. Kokovtsov, known as a liberal paternalist, as chair. The commission consisted of prominent state officials, representatives of various business groups, and members of the reform-minded intelligentsia. It also invited representatives of local governments (*zemstvo* and *duma*), members of the factory inspectorate, factory-law specialists, and members of the working class to offer their opinion about its proposals.

The commission produced drafts of new labor-legislation provisions that were published in *Torgovo-Pomyshlennaia Gazeta* ("Commerce and Industry Gazette") and widely publicized in other periodical publications. Although it is not clear whether working-class representatives formally participated in the resulting discussion, business and scholarly groups sent in suggestions to the commission. Retaining the laws of 1882 as the basis, the new legislative proposal tried to impose additional regulations on child labor. These included a maximum workday of ten hours for juveniles between ages fifteen and seventeen and seventeen nonworking holidays in addition to Sundays.[42] The draft contained five new legislative propositions. They included provisions on the workday and its divisions, on medical care for industrial workers, and on state health insurance funds. Two provisions aimed at revising existing laws that outlawed strikes and workers' associations. The provisions on medical services for workers contained more specific stipulations for implementation than the earlier acts.

The most controversial proposition focused on limiting the workday to eight or ten hours, depending on the industry and the character of work. Most entrepreneurs objected vociferously to this proposition. They pointed out that many Russian industries already had a ten-hour workday and that most other countries had no such universal regulations. The 1901 British act limited the working week to fifty-five and a half hours only for women in the textile industry and to sixty hours in other industries. The French legislation of 1892 imposed the ten-hour day for juvenile workers and women and extended this provision to all workers only in 1900.[43] Most other industrialized nations had far fewer such regulations.

Consequently, the Kokovtsov's commission's proposition regarding the workday did not come into force. The standard workday remained eleven and a half hours, the norm introduced by the 1897 law.[44] The propositions on strikes and workers' unions, however, were actually formulated as laws and enacted. With some restrictions, the laws of 1905–6 legalized strikes and provided a basis for the organization of workers' unions and cooperatives "aimed at pursuing economic interests and improving labor conditions of their members."[45] Restrictions on strikes applied to types of industry and businesses defined as of "vital importance to the nation," such as transportation, telegraphs, the postal service, banking, and so on. These last statutes allowed for the expansion of the legal workers' movement often noted in histories of the post-1905 era. Although strikes were legalized in 1906, it must be noted that workers actively utilized this form of labor protest well before the 1906 legislation. With few exceptions, strikes were resolved peacefully, by means of negotiation and compromise between the involved parties.[46]

Although the commission's proposition about insurance did not come into force at once, it provided a foundation for the 1912 insurance law. The 1912 law, with its over five hundred articles, established compulsory medical insurance and medical funds for all industrial workers and financial compensation for workers and members of their families for work-related accidents, injury, or death. The law instituted elected insurance boards, which administered funds collected from compulsory contributions made by employers and workers. The implementation of this law proceeded quite expeditiously. By June 1914, Moscow province alone had 344 insurance boards, representing 370,000 workers. By the end of 1915, fully 77 percent of Moscow factory workers belonged to insurance funds. There were similar results in other major industrial centers of Russia. According to G. A. Arutiunov, a historian of the workers' movement, by June 1914 over 2,800 insurance boards representing over two million workers, including children, had been established throughout imperial Russia.[47] Labor unions, worker-oriented cooperatives, and a host of other worker associations underwent a similar expansion, as often noted in historical literature of the era.

All of this activity was capped in 1913, when for the first time, and entirely unnoticed in the historical literature, all existing labor laws were collected into a single volume, the Factory Law Code—Russia's first uniform and comprehensive law on industrial labor.[48] All these laws affected the lives of not only adult workers but millions of children and juveniles who still worked in factories and other production establishments.

In order to facilitate the laws' implementation and aid the factory inspectors' oversight, the government provided broad publicity for the expanded labor and welfare laws. Laws were published in inexpensive single volumes affordable to most people. During the late imperial decades, several such publications addressed factory and child labor laws and explained their significance. To make them comprehensible for common and semiliterate people, these publications used plain, simple language and sometimes appeared in editions printed in larger letters. In this case, other segments of society, including, for its own reasons, the radical movement, also joined in the effort to publicize the new labor laws. A 1915 publication about child labor laws, "Our Laws on Protection of Child Factory Labor: A Common Guide," edited by M. Balabanov, provides an interesting example. The publication was divided into sections that addressed specific aspects of the child labor regulations. Each section began with large-font titles and simply written, clear statements such as, "Children under the age of twelve are banned from employment," "Children are banned from nighttime work," "Children are prohibited from work on holidays," "Children are banned from employment in harmful occupations," and so on.[49] Many periodicals of the period devoted considerable space to factory legislation, providing the issue with forums for broad public discussion. In fact, ever since the introduction of the 1882 law, numerous periodicals, including newspapers, regularly published articles about factory labor legislation and all manner of related issues.[50] The government's publications signify the tsarist state's attempt to improve and ease its dialogue with workers.

How does this process of the introduction of labor protective laws in Russia fit with those in other industrializing European countries? Although Russian industrialization began somewhat later than in several other countries of northern and western Europe, the pace and timing of the labor laws' introduction in Russia nevertheless conformed to the general European pattern. In most industrializing countries, the most decisive laws regarding child and women's labor, the workday, and the institution of factory inspectors appeared during the later decades of jthe nineteenth century. For example, as already mentioned, in England the 1833 legislation that forbade the employment of children under age nine and introduced factory inspectors in the textile industry was extended to all industries only in 1867. The 1844 Factory Act limited the working week for children under age thirteen to thirty-six hours. France banned the full-time industrial employment of children under age twelve and instituted factory inspectors in 1874. (The French law still allowed part-time employment for children between ten and twelve years of age in some exceptional cases.) Belgium introduced its first law regarding child labor and factory inspectors in 1889. (A Belgian law of 1884 prohibited boys under age twelve and girls under age fourteen from work underground in the mining industry.) An 1889 law restricted children's and women's employment and established factory inspectors in the Netherlands.[51]

Elsewhere in Europe, as in Russia, the timing of the introduction of the freedom to strike, labor unions, and social insurance laws varied, but in general, it occurred during the late nineteenth and early twentieth centuries. Although Britain had a long history of workers' unions, the first law that fully protected the country's trade unions from illegality appeared in 1871. Germany pioneered in the introduction of work-related illness and accident-compensation laws in 1883 and 1884, partly as a response to the growing socialist movement. Nonetheless, in 1886, Prussian police restricted strikes and in 1901 prohibited them. In 1897, Britain introduced the Workmen Compensation Act. In 1916, Denmark established industrial accident compensation for workers. Although well before the outbreak of World War I most European nations had abolished penal sanctions against strikes and trade unions, during the war some countries such as Britain outlawed strikes and increased the harshness of government policies toward the workers' movement. During the years before the outbreak of the war, Germany took a much stricter position toward union activism, especially strikes, than it had in the past. After the war, as a response to the rise of the socialist movement among workers, almost all European nations at one point or another introduced the eight-hour workday and unemployment compensation on their way to the creation of modern welfare states.[52]

The timing of the introduction of labor-related laws in Russia, as well as their substance, renders problematic the notion of Russian "backwardness" emphasized by some contemporaries in Russia and by many commentators to this day. In an 1882 issue of *Vestnik Evropy* ("Messenger of Europe"), as a response to the 1882 law, an editorial remarked that Western countries like Britain, France, and Germany "far surpassed [Russia] on the path toward rational factory legislation."[53] This was one of many analogous remarks. The contemporary emphasis on Russia's "lagging behind" has influenced many scholars of modern Russian history to utilize the concept of backwardness as a powerful—indeed, all-embracing—methodological paradigm for understanding and explaining Russia's past. However, the above exploration of Russian labor laws and their implementation, as well as the process of lawmaking, suggests that the contemporary remarks of Russian commentators exaggerated the actual situation. Contemporary overemphasis on "backwardness" seems to have distorted Russian reality. This tendency probably reflected the desire of some political groups within Russia, such as the famous Westernizers, to make a strong rhetorical case for speeding up the process of Russian industrial and social development, which de facto was already well under way during the late imperial decades. The notion of "backwardness" served as a discursive strategy in the contemporary debates of the day and should not be taken uncritically at face value by historians today. Doubtless, as regards aspects of industrialization and labor protection, Russia lagged somewhat behind several of the most advanced nations, such as Germany, France, and Great Britain. Even so, the gap was smaller than usually believed and, furthermore, did not apply to Russia's relative position with regard to many other industrializing nations of that era.

The Impact of the Child Labor Laws on Children's Employment

How important and effective were the laws that regulated child labor? Was their introduction significant for the lives of working children? Historians still debate the effectiveness and importance of child labor laws. Indeed, this question may be too difficult to answer definitively at this stage. Most recent studies of child labor argue that labor protection laws appeared in most countries at a time when most of their provisions had already lost their importance. For instance, Clark Nardinelli points out that restrictions on children's employment were introduced in the textile industry in 1833, when the number of employed children under the age of nine had already declined.[54] Other recent scholars emphasize that labor protective laws for women were ineffective and gender biased—primarily concerned with the protection of women as mothers, not as workers, and for the most part aimed at eliminating women from production and confining them to the private, domestic sphere.[55] By contrast, some early scholars of child labor have suggested, to the contrary, that child labor laws decreased children's employment in factories, which ultimately reflected these laws' significance.[56]

Soviet historians have devoted only sporadic attention to labor laws. Since the late 1920s, no specific study of labor laws has appeared. The voluminous literature on the labor movement and workers' unrest created the impression, despite the absence of systematic research, that tsarist labor laws were either ineffective or simply did not exist. Not surprisingly, V. I. Lenin tended to evaluate tsarist labor laws as useless. In suggesting late tsarist labor legislation's futility, Lenin compared it to carrying water in a sieve.[57] Soviet scholars embraced Lenin's assumptions. Nevertheless, as previous sections of this study suggest, the introduction of labor-related legislation occurred in Russia at a time when many industrializing countries of Europe were introducing similar legislation (and when others had not done so at all). Thus, given the relative lateness of Russia's heavy industrialization, Russian labor laws could be argued to have come into existence and to have exercised a crucial effect at the very time when the employment of children was reaching its height. In Europe, however, owing to the pace of industrialization, these laws were absent or initially highly ineffective when the need for labor protection laws was at its height.

In contrast to Soviet scholars' oversimplified accounts and evaluations, contemporary observers and factory inspectors offered a quite complex and nuanced picture of the effects of the laws of 1882 and later. They too certainly acknowledged the difficulty of the laws' implementation. Yet factory inspectors also noted the general decline of children's employment in industries, the reduction of working hours, and some improvement in working conditions for factory children. For instance, in his 1885 report, Chief Factory Inspector Ia. T. Mikhailovskii remarked that those inspectors "who visited the same factories before and after 1884 could not miss the pleasant change that had occurred in conditions for working children. Children had become more energetic, their faces fresher, ... [conditions] that had almost not existed [before the new laws]."[58] Whether this remarkable change really occurred or not, all factory inspectors clearly recognized the importance of factory labor protective laws and tried to facilitate their implementation.

To be sure, the laws of 1882, 1884, and 1885 initially regulated child labor only in private businesses that, as noted, used certain kinds of technology and employed at least sixteen workers. Furthermore, these laws applied only to European Russia. Although businesses affected by the new regulations employed hundreds of thousands of children, the laws did not address labor in agriculture, domestic services, and small artisan workshops that also employed many children. For example, within one industry, the authority of factory inspectors extended only to large matting mills and did not cover numerous small matting enterprises that did not have steam-powered technologies or that employed fewer than sixteen workers. Many children nevertheless worked in such enterprises.[59] The state tried to resolve this issue by the introduction of the 1897 law. The law extended labor regulations and factory inspections to rural workshops, thus at least in theory placing more employed children under state control and protection (in reality the state's ability to oversee the huge number of small rural enterprises was limited). In addition, as mentioned, specific labor laws regulated child labor in state enterprises and mines.

Nonetheless, the labor of children working in agriculture and domestic service, where labor conditions could be as harsh as in industries, still remained entirely unregulated and unprotected. Although coherent statistics on children who worked in agriculture or domestic service are nonexistent, many contemporary periodicals and literary publications implicitly suggest that the percentage of children employed in these sectors reached high levels. Furthermore, after the introduction of the 1882 law, many children under age twelve from poor families in all likelihood shifted to agriculture and domestic service out of sheer necessity. Thus, the fact that the child labor laws did not address all employed children probably constitutes their greatest single weakness.

Regardless, the introduction of a legal basis for a system of state control over factory labor was one of the most notable accomplishments of the 1882 law. Factory inspectors began to gather systematic data on children's employment, education, and working and living conditions in industries. They also gathered important general information on private businesses located in their factory districts. Inspectors revealed the existence of a significant number of businesses that had not previously been reflected in any statistical or police registers. For example, in 1885, an assistant inspector of the Kazan' district found in the city of Orenburg twelve factories of which the local statistical committee had no record, local fiscal authorities unaware of even their existence. Such "hidden" unregistered businesses were discovered in other industrial districts as well.[60] This information was crucial in helping the state create a more accurate picture of private industry and define more precise taxation policies. In this regard, in addition to their major responsibility—overseeing labor—inspectors supervised the accuracy of payments of certain taxes on businesses.[61]

As mentioned, during 1882–85, factory inspectors visited about 5,000 enterprises, or about 19 percent of all businesses that fell under their jurisdiction.[62] At first glance, this number may appear less significant than it actually was. These 5,000 enterprises were located in European Russia, spread over a territory of 4 million square kilometers. The St. Petersburg district was territorially the largest. It included seven northern and Baltic provinces and covered over 1.14 million square miles. Other big factory districts were Moscow, Vladimir, and Kazan'. The Moscow district included about 7,000 businesses. The vastness of the empire and its inadequate transportation system presented the biggest problem facing factory inspectors. In order to inspect a single factory, they often had to travel large distances. Factory inspectors complained that by law, they and their assistants were obliged to visit all businesses and therefore could not inspect any single business more than once over a considerable period, although many enterprises required additional follow-up visits. Thus, although factory inspectors worked quite effectively, as noted by many contemporary periodicals, they could not possibly cover all factories. For example, in 1885, one Moscow district inspector with his assistant actually inspected only 460 factories out of 7,000. The Vladimir district inspector visited 292 businesses out of the 4,065 that came under his jurisdiction.[63] The increase in the number of inspectors during the 1890s, however, brought more effective supervision of factory labor.

As noted, complete reports of the inspectors from all nine districts appeared in 1885. The first reports came from the Moscow, Vladimir, and St. Petersburg districts in 1883. Between 1883 and 1917, factory inspectors compiled and published their annual surveys, which even today are among the most valuable and comprehensive surviving sources on late imperial factory labor. Although these surveys do not reflect child labor in agriculture, domestic service, state enterprises, mines, and many small artisan workshops, they nevertheless suggest the dynamics of children's employment in private industries (see chapter 2).

Most important, the inspectors' surveys show that after the enactment of the 1882 law, the number of children working in industries rapidly decreased. For example, the inspector of the Vladimir district, Dr. P. A. Peskov, reported that in 1882–83, children under age fifteen accounted for 10.38 percent of industrial workers of Vladimir province. In 1885, the number of employed children below fifteen fell to 3.80 percent of the workforce. Overall, in the more inclusive Vladimir factory district, of the 97,756 workers employed in the 292 factories that Peskov visited in 1885, 6,049 were children. This equaled 6.05 percent, a figure that, in Peskov's words, was "significantly less than before the introduction of the law."[64] In Kostroma province, before 1884 there were 1,735 children under age fifteen working in the province's industries. After the law was enacted, there remained only 695—less than half the previous number. In the Kharkov factory district, the number of children under age fifteen decreased from 3,325 before the law's enactment to 1,425 in 1885.[65]

Inspectors also noted the rapid decline of children's employment in particular industries. For example, before 1884 about 24.0 percent of textile workers were children, whereas in 1885 children accounted for only

5.5 percent. Child labor also declined dramatically in chemical plants, where before 1884 children made up 14.5 percent of the industry's workers and after 1884 the number decreased to 0.3 percent. Inspectors noted that the decline in children's employment was especially significant at large, technologically advanced enterprises.[66]

How to account for such a rapid decline of child labor after the introduction of the 1882 law? According to factory inspectors, the 1882 law, when industrialists learned about its provisions, resulted in the dismissal of a great number of children from factories. The employers fired not only children who according to the law could not be employed but even older ones (twelve and above) whose employment was allowed. Peskov observed that "with the introduction of the law [many] owners dismissed children from their factories." Some owners fired children as "a demonstrative act, because they did not want to allow factory inspectors [to visit] their businesses." Other technologically advanced enterprises really had no acute need for child labor and even if they employed children did so only in very limited numbers as an exception.[67] The latter used child labor in auxiliary tasks and could easily replace it with the labor of adults.

Another important factor that stimulated the immediate decline of child labor in industries after 1884 was the abovementioned general economic recession during the early 1880s, a factor also stressed by many factory inspectors in their reports. As a result of overproduction, numerous factories closed or laid off many thousands of workers. Without great difficulty, factory owners first dismissed working children. By the end of the 1880s, however, when the crisis was over and the economy had begun to recuperate, the number of child workers under age fifteen had increased to 7.7 percent, less than during the late 1870s but more than during the years of economic crisis.[68] As mentioned, when the economy again expanded, employers began to attack labor laws and the factory inspectorate, a phenomenon that, incidentally, suggests the likely effectiveness of the laws and the factory inspectors.

The statistical decline in children's employment, however, may have been offset somewhat by evasions of the law that occurred with particular intensity after the economic crisis came to an end. Factory inspectors complained that child labor regulations were difficult to enforce because employers often evaded them with the complicity of parents and children themselves. As noted earlier (see chapter 2), child workers, who came mostly from impoverished working and peasant families, tried to hire themselves out in order to sustain their own lives and, quite often, provide some support for their families. In order to obtain employment, underage children concealed their real ages and claimed to be older than they were. One contemporary account of child workers in mining stated that "most of [the children] are hardly even thirteen; ... many seem to be eleven. But if you ask one of them, 'How old are you?' to your astonishment, he will answer, 'Fifteen.' This [occurs] with the knowledge of the mine administration ... and it is not in the interest of the boy himself to reveal his true [age]—he can lose the job."[69] According to inspector Peskov, "One cannot fully rely on the age information in children's documents, a fact about which I personally became convinced.... Even entrepreneurs themselves share the opinion that the identification information about ages is inaccurate.... According to their documents, some children were thirteen or fourteen years of age, but their external appearance and physical development suggested that they were no more than ten."[70] Local authorities sometimes issued documents that stated the age necessary for factory employment, even if this involved adding a couple of years. They often did so with the agreement and for the benefit of parents who wanted to send their offspring to factories. One worker, Aleksei Buzinov, later wrote in his memoirs that when his father died, his family experienced considerable hardship. He was only eleven years old, and his mother had great difficulty finding employment for him, although they desperately needed supplementary income. Through one of her deceased husband's peer-workers, she was able to obtain a false birth certificate for her son that added two years to his actual age. With this false identification, he was employed in his father's plant as an apprentice.[71]

Factory inspectors were well aware of these practices and usually did not take the ages stated in children's identification documents for granted. They tried to estimate a child's age by his or her appearance and also asked the children themselves about their ages. The responses were not always exact, because in some cases children did not even know or in other cases wanted to conceal their real ages. Peskov reported that once, after he had finished his interviews with working children in a calico-printing factory, "one embarrassed-looking boy suddenly returned and stated that he was not thirteen years old as he had said but only eleven." When Peskov asked him why he wanted to conceal his age, the boy replied that his overseer had told him to do so. In addition, during inspectors' visits, some owners tried to hide employed children by sending them to places within factories where inspectors could not have access, thus corrupting the accuracy of data on children's employment.[72]

Even in 1900, some eighteen years after the introduction of the 1882 law, inspectors disclosed violations regarding the employment of children under the age of twelve. For example, inspections revealed that in 1900 eight factories in the St. Petersburg industrial district, three factories in the Moscow district, and ten factories in the Warsaw district used the labor of children under age twelve. Similar violations were found in other factory districts.[73] According to police records, employers who transgressed the law were subject to fines as high as one thousand rubles, although most fines were about one hundred rubles.[74] This phenomenon—factory inspectors regularly finding violations of the laws regarding the age of child laborers—suggests the probable overall reliability of the factory inspectorate's data on the ages of child workers. Inspectors usually observed and talked to children in person and registered them in the appropriate age group according to direct observation, rather than according to the factory's data. This allowed inspectors to disclose cases of legal transgression and report them to the police.

Even conceding possible evasions, taking a long-term perspective, during the three decades before World War I, the employment of children in industry gradually declined. As mentioned, in 1883, the year before the introduction of the law, children between the ages of twelve and fifteen (*maloletki*) accounted for about 10.0 percent of factory workers in Russia. By mid-1885, this figure had fallen to 3.9 percent. This tendency continued until the outbreak of World War I. The number of children between ages twelve and fifteen decreased, whereas the number of juveniles from ages fifteen to seventeen slightly increased. In 1901, working children between ages twelve and fifteen accounted for 2.0 percent of industrial workers, juveniles between ages fifteen and seventeen for 8.6 percent. In 1905, *maloletki* comprised 1.4 percent of workers, juveniles 9.0 percent. In 1913, industrial labor consisted of 1.6 percent *maloletki* and 8.9 percent juveniles.[75] Thus, over a period of thirty years, the number of factory children under age twelve had fallen to insignificance, and the number of child workers (*maloletki*) had fallen from 10 percent to less than 2 percent.

Available data from individual factories confirms this picture of a significant decline in children's employment. By 1907, in the Putilov plant, one of the largest metallurgical enterprises in St. Petersburg and all of Russia, working teenagers accounted for only 1.3 percent of the workforce. The St. Petersburg Tentelev Chemical Plant did not employ children at all. In metallurgical and chemical industries, perhaps the most hazardous to children, children's employment declined significantly after the introduction of the 1882 law. In certain other plants, however, the percentage of children still remained high. For example, the Torkovichi Glass Mill employed 238 children between ages twelve and fifteen, fully 43 percent of the mill's workers. Most of these children were recorded as apprentices.[76] Nonetheless, employment in agriculture and domestic services aside, it is clear that employment of children below age fifteen in factories was disappearing in late tsarist Russia. A number of early Soviet memoirists and autobiographers who entered the workforce after the 1891–92 period recalled that they began their employment between the ages of eleven and fifteen, a factor that also indirectly suggests a decline in very young children's employment.[77]

How did the child labor law affect labor conditions for children working in industries? This question quickly became controversial and indeed, toward the beginning of the twentieth century, ideologically loaded and politicized. Various political parties used labor issues to attack the government and appear the best protectors of workers' interests. These groups recognized no improvements brought about by the labor laws and tended to accentuate the worst cases in anything involving factory labor. According to some radical socialist periodicals, the conditions of working children and juveniles improved little in comparison to previous decades before the introduction of labor protection laws. For instance, *Iskra* and *Proletarii*, two famous Bolshevik newspapers, cited examples of working and living conditions of children employed at the Filipov Candy Factory in Moscow. Children received five rubles a month, room, and board. Their workday lasted eleven and a half hours during the daytime and ten hours at night. Children lived on the top floor of the factory building in a room without air circulation that housed about three hundred people. Beds were set up in pairs, each pair accommodating three or even four people.[78] Although such cases may have accurately reflected the reality of working conditions at particular enterprises, they by no mean represent overall reality.

Indeed, evidence regarding labor conditions is much too diverse and fragmentary to allow for strict conclusions about whether or not they were bad or how bad or good they were. As noted, contemporaries, including factory inspectors, observed that after the introduction of the 1882 law, labor conditions for working children witnessed relative improvement by the end of the nineteenth century. Factory inspectors reported that businesses, when required, introduced safety measures, such as covering the moving parts of machines and steam engines, replacing wooden stairs with cast iron, improving air circulation, and so on. Some businesses reorganized the placement of machines and equipment on shop floors in order to provide wider spaces and passages for safety reasons.[79] The abovementioned Aleksei Buzinov, the worker who started his employment at age eleven by using false identification, recollected in his memoirs that his experiences as a smithy apprentice were varied. Although his tasks were "laborious," in the main, he later insisted, he learned to appreciate the value of labor. Aside from their casual teasing, adult workers watched over him and taught him the blacksmithing craft. When he gained the necessary skills, his work became a source of satisfaction for him. Out of this satisfaction, he learned "to appreciate things" he had produced with his own hands.[80] This young individual, like many others, achieved a sense of pride and self-worth through his productive skills. This should remind us that for many working-class children and youths, entry into productive labor was a desirable part of life, a rite of passage toward adulthood. The exploitation of children was a black mark on all industrializing societies, but many laboring children and their parents doubtless saw the matter differently.

In one important area, real improvement occurred. Late imperial statistics illustrate a definite decline in work-related accidents among workers, which also signifies improvements in labor conditions. The number of work-related accidents among workers under age seventeen in fact decreased dramatically. According to the 1894 data from Vladimir province, where surveys covered 75,522 workers (including 6,179 children and juveniles under age seventeen), work-related accidents requiring a physician's attention occurred to 1,904 workers (2.5 percent of the workforce), including children and juveniles. Injured children and juveniles accounted for 224 of these, or 3.6 percent of working children and 11.6 percent of all injured workers. The number of injured adults was 1,680, which accounted for 2.4 percent of working adults and 88.4 percent of all injured workers.[81] Although the proportion of injured children was still relatively higher than that of adults, it is clear that the overall number of work-related injuries had declined dramatically since 1884. In contrast, before the enactment of the 1882 law, more than 50 percent of all work-related accidents occurred to working children (see chapter 2).

In addition, inspectors noted that the law affected the actual workday for children. Before the law's enactment, the regular workday for children lasted from 12 to 13 and even more hours. After 1884, a child's workday approached 6 or 8 hours, depending on the type of labor organization. According to a mass of 1904 data that covered 1,366,000 workers, the workday averaged 10.7 hours for adult males and 10.4 hours for women and children between ages fifteen and seventeen. This was less than the norm set up by the 1897 law. Children under age fifteen worked 7.6 hours. In 1913, the maximum workday lasted 11.5 hours. Some historians point out that these data came from official reports produced by factory administrations interested in "underestimating" the length of the workday. In this version, the actual workday might have been somewhat higher.[82] The data issued by factory administrations are, however, supported by the reports of factory inspectors. This data suggests that teenagers between ages fifteen and seventeen on average worked 9.83 hours a day, those under fifteen 7.9 hours.[83]

It is also an indisputable fact that the decline in the workday for children directly—and negatively—affected their salaries. In most cases, children's wages decreased proportionally, relative to the reduction in working hours. According to factory inspectors, with the decrease of working hours from twelve to eight, children's wages were lowered by one-third; when the workday was reduced to six hours, children began to receive half of their previous wage. In the Kiev industrial district, children sometimes did not receive any wages but worked for room and board.[84] Factory inspectors suggested, however, that the reduction of working hours in fact led to an increase in children's hourly productivity. Obviously, children worked shorter hours and were less overworked, as a consequence of which they presumably could work more effectively and produce more per hour. Regardless, children's increased productivity rarely had a positive effect on their wages. Only a few employers increased children's wage rates when they realized that their productivity had risen.[85]

Even so, existing data on wages suggest a general rise in adult and juvenile wages. Throughout the empire, in 1905 the worker's average annual salary was about 235 rubles, in 1910 it was 246 rubles, and by 1913 it had further increased to 264 rubles. The highest average salaries for workers were in the St. Petersburg industrial district, where in 1913 workers got about 339 rubles, whereas the lowest average wage of 196 rubles was recorded in the Kiev district. In 1900 St. Petersburg district workers on average received 265 rubles, whereas in the Kiev district workers' average wages were 133 rubles.[86] Regardless, the rise in the prices for daily necessities led some observers to point out that increases in workers' wages were partially consumed by inflation.[87] Data on workers' expenses, however, suggest that on average, in the late nineteenth century, workers confronted roughly the same outlays for foodstuffs as they had before 1884 (see chapter 2 for data on workers' food expenditures during the 1870s). During the early twentieth century, an average adult worker spent monthly from 4 to 5 rubles for food, whereas children's expenses ranged from 2.25 to 4 rubles a month. Workers' expenses depended on their wages. Those who received higher wages tended to spend more on food. Dement'ev estimated that an average working family spent about 58 percent of its income on food, with variations depending on the size of the family.[88]

Nonetheless, the outbreak of World War I created new realities, which led to some negative effects on ameliorations brought about by the labor laws. With the beginning of the war, many men left the factories for armed service. Consequently, industries faced a great demand for labor. By 1916, the demand for workers greatly surpassed the labor supply. The government introduced detailed regulations that allowed women and children (between ages twelve and fifteen) to labor in those industries and occupations where previously they had faced prohibitions, such as metallurgy and mining. The 1915 statute also permitted underground work for women and children. At this time, many women and children entered the industrial labor force. If in 1913 industrial workforces consisted of 13.9 percent teenage workers (between ages twelve and seventeen), in 1916, as an immediate impact of the severe labor shortages during the war, this number increased to 21.0 percent.[89] Just before the February 1917 revolution, factory inspectors recorded 49,956 child workers between ages twelve and fourteen (2.4 percent of the workforce) and 242,866 juveniles between ages fourteen and sixteen (11.6 percent), in a total industrial workforce of 2,093,860 persons in the private industries covered. In 1913, children and teenagers of these ages had accounted for respectively 1.4 and 9.7 percent of factory workers.[90]

After the February 1917 revolution, the provisional government attempted to resolve the child labor issue, which the war had exacerbated. Child labor was one of the most vigorously debated questions of the newly created Ministry of Labor. In March 1917, the provisional government abolished the 1915 statute that had allowed military-oriented mining and metallurgical industries to use the labor of children and women, including for underground work. The law of August 1917 abolished nighttime work for juveniles under age seventeen and for women in all industries. For the duration of the war, the labor minister, however, retained the right to allow nighttime work for women and children, with the agreement of the minister of trade and industry. Regarding child labor in general, the provisional government retained all previous provisions of the 1913 Code on Industrial Labor.[91]

The October 1917 revolution and its aftermath produced new social and economic realities that altered the nature and perceptions of child labor. After the October revolution, Russia faced civil wars. The well-known national economic collapse threw many hundreds of thousands of workers into unemployment. The number of children under age fifteen employed in factories declined dramatically. By September 1918, teenagers between ages fifteen and seventeen accounted for 13.1 percent of the factory workforce; by July 1919 this figure had further dropped, to 8.5 percent. In general, the period of War Communism (1918–21) left behind little statistical evidence. One source suggests that during 1918, unemployment reached 1,500,000, a figure that doubtless impacted children as well as adults.[92] During the years after 1918, Russia faced the tremendous social problem that contemporaries called *besprizornost'* (children's homelessness and neglect), which involved several million children.[93] With the collapse of social institutions that provided care and education for children, millions found themselves on the streets. But that is another story.

In summary, although factory labor laws lagged behind the pace of children's involvement in industrial labor and therefore had little impact on the generation of children who first experienced industrialization, Russia, like other countries, did introduce laws about child labor. Russia's labor laws came at the very peak of industrial development, perhaps somewhat in advance of the pace in most industrializing countries. These laws improved conditions for children and certainly had the potential for improving the well-being of future generations of children in Russia. Child labor in industries became subject to state control and protection. Factory inspectors gathered important data on factory labor and supervised children's employment. In addition, the laws recognized education as a priority of childhood and as a desirable alternative to factory labor. Finally, and most important, industries could no longer regard very young children as a source of labor and had to seek production methods, technologies, and organization of labor that would end their dependence on young children's employment.

The Education of Factory Children

Compulsory education of children employed in factories was another significant aspect of late imperial labor law. A few words about education in imperial Russia will help provide a context for the issue of the education of working children. Before the reforms of the 1860s, elementary education for children of all social estates was provided in district schools (*volostnye* and *uezdnye shkoly*), elementary schools for peasant children, elementary schools of the mining industry, and orphanages. However, the vast majority of children in Russia, especially serf children, remained outside these schools. Peasant schools were usually limited to state and royal family villages and were simply nonexistent in serf communes. In 1836, there were only sixty-five peasant schools, whereas by the mid-1850s their number had increased to twenty-five hundred.[94] Elementary schools in serf villages were solitary exceptions. The evidence on such schools is extremely limited. A few serf children received an elementary education privately with priests, retired soldiers, or village communal scribes. For example, the former serf Savva Purlevskii recalled in his memoirs that he studied basic literacy and calculus with the local priest and then with his father. When Purlevskii grew up and became a bailiff in the late 1820s, his village commune and the landlord founded a school for village children.[95] Nevertheless, the majority of serf and numerous state peasant children remained illiterate or barely literate.

In addition to these scarce educational opportunities, some children could receive an education at factory schools. The history of factory schools in Russia perhaps dates back to the early nineteenth century. During the first half of the nineteenth century, the growth of new mechanized industries, with their elaborate technologies, created a new demand for educated workers. New complicated machines required not only workers with elementary literacy but those capable of mastering new techniques. Deeply concerned about qualified workers at a time when the government restricted education for lower social estates to elementary schooling, some entrepreneurs, of their own private initiative, began to establish factory schools, technical schools, Sunday schools, and schools for teenage workers.[96] Beyond promoting education among workers, some owners saw these schools as a means of "social control," creating loyal, disciplined individuals.[97] Nevertheless, before the 1884 law, the effort to spread education among working children remained highly sporadic, depended on the employers' goodwill, and was usually limited to a few large enterprises.

The history of early trade and technical schools founded by the brothers Timofei Prokhorov and Konstantin Prokhorov (co-owners of the famous Three Mountains Factory in Moscow) is a notable example of entrepreneurial endeavors to promote education among children. The first Prokhorov school opened in 1816, for two hundred children of the factory's workers (most of whom were peasant-migrants) and the Moscow poor. In 1833, Timofei Prokhorov opened another school for both children and adults. Education in both the Prokhorov schools was free. To maintain their schools, in 1840 the Prokhorovs spent about 17,000 banknote rubles and, in 1842, 25,845 rubles—remarkable sums by contemporary standards.[98]

These were, however, exceptions. In the 1840s and 1850s, there were only thirty-four factory schools in Moscow province, including sixteen factory schools in Moscow, with over 1,000 students.[99] In addition to these few factory schools, in 1843 ten Sunday factory schools opened in Moscow province, with 1,050 students. A modest number of factory schools existed in other provinces of imperial Russia. Although all these educational establishments were private, the government attempted to regulate their general curriculum. Students of these schools received an education in industrial technology, industrial chemistry, factory management, mechanical drawing, machine construction, accounting, and other technical and financial disciplines, as well as in general subjects such as religion and calculus.[100]

The significance of these educational establishments was that they were open to children regardless of their social background and gender. According to a report for 1844, "the major part of students of private factory schools belongs to the peasant estate [including serfs] and less that one seventh are from petty townspeople [*meshchane*]."[101] The Finance Ministry's technical drawing schools represent another interesting example. Among 874 students of the schools, 109 were serf children, 131 were children of peasants of other categories, 31 were nobles' children, 56 came from the families of high military officials, 18 were from the clergy, 2 were from the state bureaucracy, 14 were orphans, 2 were honorary citizens' children, 52 were the children of "people of various ranks [*raznochintsy*]," and 467 were from other social estates—mostly townspeople.[102] Teenage girls were among the classmates at some factory schools: 80 female students attended the Prokhorov, Guchkov, and Roshfor factory schools.[103] These fragmentary statistics hardly represent the full number of working children who received an education, a phenomenon that should be neither exaggerated nor ignored. Of course, children of the nobility, clergy, and townspeople could receive an education at other schools or from private tutors. Factory and Sunday schools were particularly crucial for working children and children from the lower social orders because these offered their only chance to get an education.

The 1860s, which left their mark on Russian history as the Period of Great Reforms, brought significant changes to the education of the lower social orders. During the 1870s, numerous *zemstvo* schools opened their doors to peasant children.[104] These significant efforts in the schooling of rural children were, however, undercut by increasing peasant migration to urban and industrial areas. Those children who moved from their villages seeking factory employment could no longer go to their rural *zemstvo* schools. Furthermore, having migrated to a city and taken a job, many children in fact lost the opportunity to receive any education at all and remained illiterate. Factory schools that working children could attend existed only in some state and large private businesses, whereas local boarding schools were often situated far away from factory districts and were not easily accessible for factory children.

At the same time, as they undertook factory employment, children could hardly find time to attend even nearby factory or district schools. In most cases recorded by governmental agencies, most rural children who had attended local schools in the countryside were no longer capable of doing so after they moved to cities and took factory employment. Chapter 2 cites the example of a twelve-year-old boy who, before his move to the city and factory employment, went to a local village school but who, after he entered the factory, where he worked twelve hours a day, could no longer continue his schooling.[105] This was the case for most working children. Chief Factory Inspector Mikhailovskii wrote in 1885 that before the 1882 and 1884 laws, it was impossible to require factory children to attend schools after twelve hours of work. He remarked that "in these circumstances, education would be more deleterious than useful.... It would lead to complete exhaustion of [the child's] immature body." Inspectors noted that students at factory schools were mostly local children and children of workers who did not work. Working children often attended their enterprise's schools, if they were even available, irregularly.[106]

The lack of opportunities for employed children to receive an education had a direct impact on their literacy rates. By 1885, factory inspectors had interviewed about 15,300 working children and found that 5,300 of them (35 percent of the total) were literate or semiliterate (could only read) and that only 500 had received formal diplomas. The balance (65 percent) were illiterate. The highest literacy rate among working children in 1885 was recorded in the St. Petersburg industrial district, reaching 70.26 percent. The lowest proportion of literate working children was in the Kazan', Kharkov, and Vilensk (Vilna) districts and ranged from about 20 to 25 percent. Literacy rates among children employed in Moscow and other central provinces was about 30 percent.[107] In the Vladimir factory district, out of 4,965 working children, 1,508 (30 percent) were literate and semiliterate. The lowest literacy rate was among working girls. Only 265 girls working in the district were either literate or semiliterate.

This number accounted for 5.3 percent of all employed children and 14.6 percent of employed girls.[108]

The evidence from some individual factories lends support to this general tendency in literacy rates. When the Vladimir district inspector Peskov visited the Sokolovskaia Cotton Mill in 1882, he found that of the 276 factory children, only 83 (30.1 percent) were literate or semiliterate. Some children (11.6 percent) attended the mill's school, located nearby. The working day in the mill lasted for twelve hours, in two six-hour shifts. Peskov remarked that obviously, after a twelve-hour workday, children were too exhausted and could hardly attend the mill's school.[109]

The laws of 1882 and especially of 1884 constituted a significant turn in the question of the education of employed children. The laws prioritized the education of working children. For the first time in Russian history, the law obliged children employed in factories to attend at least a one-year program of elementary schooling and receive a diploma. Those children without the required education had to receive an education either before entering factory employment or during employment. In addition, the reduction of the workday for children to six or eight hours opened an opportunity for them to attend factory or local boarding schools. The number of factory schools, however, still remained low despite factory inspectors' efforts to motivate employers to build such schools. According to the author of an 1894 article in *Russkaia mysl'* and the reports of factory inspectors, in 1885 there were only 163 private factory schools.[110] By 1899 their number had increased to 446, at which point about 44,400 working children were attending these schools.[111]

Although the number of factory schools was low and they could not accommodate all employed children who needed an elementary education, during the late nineteenth century, many employers and philanthropic societies did undertake a significant effort to promote literacy among their workers. Late nineteenth- and early twentieth-century sources offer abundant evidence of employers' support for the education of their workers. For example, the Ramensk Mill founded a school that eventually educated 374 boys and 301 girls. In 1907, the owners also built structures in the village of Ramenskoe for schooling the children of local peasants. These buildings, along with ten thousand rubles, were given to the local *zemstvo* for founding a boarding school (*narodnoe uchilishche*).[112] Some mills also set up subscription libraries and organized Sunday readings for their workers. For example, the Ramensk Mill had a library with a total of 26,658 volumes, including textbooks, educational and popular literature, and periodicals.[113] The Ramensk Mill had a rather remarkable record of literacy among workers. According to the 1914 data, of the 630 recorded workers, 76.8 percent were literate, 3.2 percent semiliterate, and 12.7 percent illiterate, with the balance unknown.[114] By the 1890s, even mining industries of the distant Lena region had set up factory schools, libraries, and other facilities for mining children. Some companies also arranged theatrical performances for their workers.[115] "Children's Labor and Rest," a philanthropic society that emphasized "book learning," opened in Moscow in 1909 to promote education and improved upbringing among the city's children. Its goal was to raise children's cultural level and deflect them from the streets and crime.[116]

As a result of this educational effort, literacy rates among employed children increased significantly during the late imperial decades. According to the 1918 census, the general literacy rate among workers was 64.0 percent—44.2 percent among working women and 79.2 among working men. Literacy prevailed among young workers. Among workers between ages fifteen and nineteen, 93.6 percent were literate; among those between ages twenty and twenty-five, 88.6 percent were literate. Data from Moscow in 1913 confirms these trends. Only 45.6 percent of men and 1.9 percent of women between ages fifty-five and sixty who worked in the city's factories were literate, whereas about 90 percent of working males and 40 percent of working females between ages fifteen and twenty-five were literate. Literacy, however, depended on locality. In many areas of non-European Russia, and in some western and southern provinces of its European areas, literacy rates among workers remained significantly lower than in the central provinces and St. Petersburg.[117]

Children's Socialization and Involvement in Political Life

One of the most fascinating and almost unknown developments of the late imperial decades was the participation of children in social and political events in the empire. Most of these children, themselves often from working families, were employed in industries, and many attended factory and other schools. In addition to the new opportunities for working children to receive an education, factory labor also seemed to facilitate children's rapid involvement in social and political life. In some instances, the factory environment proved to have a corrupting influence on young individuals. Factory children worked side by side with adult workers and often resided in the same crowded quarters with unrelated adult people, where, as one historian of Russian labor put it, "people cooked, smoked, argued, chatted, and tried to rest, [and] children dashed around."[118] One eyewitness of young workers' lives in an eastern Siberian gold mine noted that "your twelve-year-old boy at the mines already smokes tobacco ... swigs down a jigger of vodka in one gulp ... and neatly washes a tray of gold."[119]

Working children learned early all aspects of adult life experience, from grievances to happiness. G. V. Plekhanov, an early Russian Marxist and theorist of political economy, observed that working children and teenagers were distinguished from their peers among the upper classes in their social independence and their material self-reliance: "Life presses upon them the struggle for existence, and this inculcates in children resourcefulness and tempering in order to avoid early destruction." Plekhanov recalled that he had met a thirteen-year-old boy, an orphan, who lived completely independently: "The boy himself settled his accounts with the factory office and knew how to balance his miniature budget."[120] Factory children also engaged in the workers' movement, actively participated in labor protests and strikes, and were often initiators of protest. Working children and teenagers also became involved in workers' associations and political parties. Most political parties in Russian had their own youth organizations.

Activists in the Russian workers' movement observed the involvement of working children. For example, when Plekhanov delivered a speech at one of the early meetings of the Land and Freedom (Zemlia i Volia) Society in St. Petersburg in 1871, he noted that the meeting attracted many school-aged children, most of whom worked in the city's factories. Plekhanov spoke under a banner upon which was written "Zemlia i Volia!" and that was held by a sixteen-year-old worker, a weaver in a textile mill.[121] Just as children assisted adult workers in the labor process, they seemed ready and willing to help workers in their political activities. Sometimes children played leading roles in labor protest.

According to numerous primary sources on the labor movement, children were frequent participants in and even initiators of demonstrations and strikes, taking their concerns into the streets. In the spring of 1878, a children's demonstration occurred in St. Petersburg. During the strike at the city's Novaia Cotton-Spinning Mill, several participants, including children, were taken to the district police. A group of children working at the mill immediately organized a demonstration and went to the police headquarters, demanding release of their coworkers. In November 1878, a children's strike broke out in the Kening Textile Mill, which employed about 200 workers, 140 of them children or teenagers between ages twelve and fifteen. The mill owners wanted the children to perform extra work in addition to their regular tasks. In protest, the children stopped work, and a strike broke out. Later the children were joined by the plant's adult workers. The factory administration, however, refused to accept the workers' demands.[122]

The record of the workers' movement during the late imperial period suggests that this was not an isolated incident. A strike broke out in 1902 in a St. Petersburg tobacco factory. This strike was started by working girls when female children who assisted adult workers refused to work for thirty kopecks a day and demanded increased pay. When refused, the girls went on strike. A strike initiated by working children also occurred at a shipyard in St. Petersburg. According to the recollections of one of the strike's participants, "a large group of boys, about 200, gathered around the factory administration. The chief master came to the boys and addressed [them] with admonitions. Instead of replying, the boys submitted a letter that demanded a raise in their wages. The master suggested that those who disagreed with the existing rates could leave the enterprise. The boys were then joined by adult workers. The strike lasted one day, and the workers' demands were fulfilled."[123] Another strike initiated by children broke out in early 1903 in the Nevskii Cotton Mill in St. Petersburg. The boys who helped adult spinners working on mule machines stopped their work and went on strike, demanding that the administration raise their wages and dismiss their overseer, a certain Nikolai Ivanov. The boys were joined by working women and later by men.[124] Similar incidents of child-worker activism took place in other areas during the late nineteenth and early twentieth centuries.[125]

Most strikes initiated by children reflected their desire for higher wages. As noted in the previous section, the introduction of the 1882 law led to a reduction of children's daily work hours and in turn decreased their wages. In some cases, children's protests were directed against adult workers who in fact employed children and paid their salaries. In these cases, children's salaries came out of workers' wages and depended on their goodwill. In some of these strikes, adult workers sided with children and demanded that factory owners raise children's wages. Most such strikes, however, would probably never have occurred had children not started them. Adult workers often seemed to support children's demands by work stoppages because they could not continue their tasks without the children's help. Self-interest, rather than charitable instinct, seems to have motivated them.

Children also assisted adult workers during acts of protest and often proved to be very handy helpers. When demonstrations took place, children often served as observers and watched out for police. When the police were in sight, the children whistled to inform demonstrators about their approach. Demonstrators then had the opportunity to disperse and hide. In some cases, children cried or made jokes in order to distract the police. When police turned toward the children, the protesters smashed street lights and windows. During one strike at Petersburg's Obukhov Plant, children helped adult workers to build barricades and resist the police.[126] One member of the Former Political Prisoners' Society, recollecting the 1905 revolutionary events in Moscow, recalled that adult workers laying down plans for a demonstration gave some coins to young boys "so that in five minutes not a single street lamp would be left burning. In two minutes, the boys had slingshots and in five—no lights were left."[127]

Demonstrations and strikes with child and teenage workers' involvement sometimes turned violent. As contemporary accounts suggest, worker protests were frequently accompanied by manifestations of misrule, such as commotion and noise, and sometimes by direct violence, including the breaking of machines, glass, windows, and so on. In some cases, children resorted to violence in order to induce other workers to participate in protests. For example, in one case, in order to get adult workers to stop work at the Morozov Cotton Mill in Tver', working children began to break windows in the factory buildings. According to a description, children and teenagers "hissed and whistled." During a strike at the Tornton Mill in St. Petersburg, working children used boiling water and stones against police. During the general strike in Odessa in May 1905, in order to bring the entire city to stop working, children rang the church bells and let the steam out of boilers. According to a police report, using these and other methods, a "band of boys compelled the shop assistants to strike."[128]

Alternatively, some demonstrations were well organized and peaceful. In such cases, before going on strike, children first presented their complaints orally or in writing.[129] As noted, by the late nineteenth century, most child and young laborers were literate and knowledgeable about factory laws. One description of a workers' protest in the Ekaterinburg Printing Mill noted that apprentices were particularly distinguished by their behavior. "The juveniles," according to this observer, "are all educated and smart and read books just like they eat a piece of a white bread cake. The employer cannot deal with them easily. If he asks them to do extra work, they refuse and refer to the law that limits their work [hours]."[130]

Contemporaries also noted that child and juvenile workers were quick to question the existing social and political order. State officials' reports noted that young workers rejected family and religious values, ignored the existing social norms, and were disobliging and disrespectful of authority. One contemporary observer wrote that industrialization led to "the decline of morality. [This is] one of the most deplorable tendencies of the past and [is] connected with the diminution of religiosity among people ... encouraged by the nihilist media."[131] The deputy minister of the interior P. D. Sviatopolk-Mirskii expressed his opinion in 1901: "In the last few years the good-natured Russian guy turned into a type of a semiliterate member of the intelligentsia who believes that it is his duty to reject religion and family, disobey laws [and] authorities, and jeer at them."[132] Police reports claimed that the "militant mood is observed only among green youth [*zelenoi molodezhi*]."

Ironically, the observations of contemporary officials find support in numerous workers' memoirs. One Jewish worker described in his memoirs how he and his coworkers broke with their religious beliefs, calling them "superstitions":

We children of poor parents hired ourselves at a bristle factory in Nevel'. There were about 150 boys. We labored about 15–17 hours a day with low wages in dirt and dust. At the end our patience had come to an end, and we went on strike.... We could not break the intractability of the owner, and the strike lasted a while. Finally, we won a 10-hour workday. Then we found other obstacles that lay outside the factory. These were our religious prejudices. We were very religious boys, so religious that at one point we donated contributions from our wages and made a present, a sacred object, a torah scroll to the owner. In the city we were exemplary boys. But when life became so unendurable, we realized that God is bad, and we scorned his help. We cast off all these religious superstitions. We began to smoke and to eat Russian sausage and pork. By doing this we caused wild hostility from fanatically religious Jews.

This experience was likely shared by thousands of working children of all religious backgrounds, as suggested by workers' memoirs published during the 1920s.[133] Having broken with religion, many factory children and juveniles entered the youth organizations of various political parties and movements. Such organizations arose in St. Petersburg and many imperial provinces. The first children's organizations appeared in the western and southern provinces of the empire. The Yugenbund

(Youth Organization) was organized under the Bund (a Jewish social-democratic organization) in 1905 and involved working children between ten and fifteen years of age from Poland and western Ukraine. In 1906, the League of Youth arose in Moscow. Other large youth associations included the Northern Union of School Youth, the South Russian Union of Youth, and Budushchnost' (Our Future).[134]

Working children directly participated in the revolutionary events of 1905. In 1905 in Dvinsk, some three hundred children went on strike. Children paraded along streets with political slogans stating "down with autocracy," "down with tyrants," and so on. During the procession, the children tossed leaflets that stated that they had "organized the demonstration not to produce a children's play but to protest against tyranny and the brutality of our government.... For freedom!"[135] Slogans that stressed political freedom and agitated against the ruling system became typical in children's demonstrations of the 1903–5 era. M. Voronin recalled meetings in 1906 at the Guzhon, a large metallurgical plant in Moscow, during which working children distracted the police at workers' meetings: "They gathered in groups on the opposite side of the workers and began to sing revolutionary songs. Cossacks rushed toward them, but they quickly disappeared.... Such meetings continued until the autumn of 1906."[136]

As noted, children engaged in the 1905 revolutionary events in Odessa. Some contemporary observers claimed that during the general strike in Odessa, working children and youths predominated. One youthful participant in those events wrote that "many people noted the numerous children who took part in the strike. Let it be! Is it not good that we proletarian children participate in this struggle? ... We, children, are exploited even more because we are more helpless."[137]

Why did children's protest occur, and why did children involve themselves in the labor movement? Even some contemporaries asked this arresting question. "How could it happen," wrote Eratov, a member of the Former Political Prisoners' Society, "that such young minors were engaged in real revolutionary activities that threatened their freedom?"[138] Unfortunately, the working children themselves left no contemporary accounts of their motivations. To a limited extent, these children's voices come down to us later as filtered through the adult memoirs of individuals who had been working children and who engaged in protest. For instance, one Moscow car-factory worker later recalled that as a child he "took part in the [1905] armed uprising in Moscow" and explained that "it was more out of mischief than conviction; I was a kid."[139]

Sociologists, psychologists, and scholars of protest explain the phenomena of protest by referring to the lack of other means to bring concerns and demands to public and state attention. But unlike general episodes of worker unrest, which could be interpreted as class conflict, children's protests may signify another conflict—that is, generational conflict. Some scholars of generational conflict suggest that this type of protest holds greater significance than class-oriented conflict. In this interpretation, class conflict is usually connected to specific issues, such as working conditions and wages, whereas generational conflict involves primarily cultural and psychological realms. It comes from "deep unconscious sources ... vague, undefined emotions which seek some issue, some cause, to which to attach themselves."[140]

The involvement of working children in political life signified the development of a new culture among the younger generation. This generation was born well after the great reforms of the 1860s and experienced the rapid tempo of the late nineteenth and early twentieth centuries' industrialization and its consequences. This generation felt itself emancipated and yearned for ever greater liberties and opportunities. This culture emphasized protest of and resistance to official values and norms, as exemplified by the state, the church, and parents. Its members emphasized political freedoms and social equality.

Some contemporaries noted the increasing generational conflict occurring during the late imperial decades. They observed that "youth felt with more strength the impassability of the gap between parents and children.... Parents joined this regime that suppresses [their] souls."[141] A working activist, recollecting his revolutionary activities in 1905, noted that when he was a minor, there was a "strong tension between youths and adults. Wherever one comes in, everywhere one sees a struggle of children and their parents. When we were doing something, we hid from our parents, and if some of the parents found out about their son or daughter being involved in the revolutionary movement, they would kick up a row.... On one occasion, the mother of one of [our] group's members [in whose house we met] threatened to report our activities to the police."[142]

Because they were young, children's and youths' hopes were high. Young people desired a better life than that of their parents. This better life was associated with broad political freedoms, a constitution, representative institutions, and—for those with radical tendencies—economic justice.[143] Despite the government's efforts to ameliorate labor conditions and promote welfare and education among workers, the cultural conflict between the expectations of a rapidly changing society and the state's stagnant political structures grew exponentially. Most citizens found themselves dissatisfied with a tsarist political system that proved itself quite incapable of dealing with the hardships caused by World War I. When the war broke out and the tsarist government showed the first signs of weakness, factory children and youth actively plunged into the revolutionary movement that ended the old regime in Russia.

Conclusion

Experience and Outcome

CHILD LABOR IN IMPERIAL RUSSIA has been an obscure page in the nation's history. Historians have usually failed to note the considerable number of children in the country's industrial workforce and, consequently, the surprisingly large role they played in Russia's industrialization. Youngsters had been involved in productive work long before modernized industries emerged in Russia. From time immemorial, children had been active in agriculture and cottage industries. They had also worked in state and manorial enterprises. Traditional societies everywhere perceived children's involvement in production of all kinds not primarily as labor but as a form of apprenticeship, an approach that certainly found broad cultural acceptance in Russia. The Russian state and society viewed apprenticeship as ethical and pragmatic, a practice necessary for preparing children for adult life. Indeed, early state legislation (pre-1800) specifically authorized the apprenticeship of children. As a result of this outlook, no early laws attempted to prohibit child labor. Thus, before industrialization, child labor did not directly reflect or result from poverty but emerged from a pedagogical social need—it was a social "good" rather than an "evil," its ultimate significance more sociocultural than economic.

Everywhere that it occurred, industrialization quickly set in motion a transformation of the meaning and purpose of child labor. When industrialization began in Russia, many children naturally entered the industrial workforce as an integral part of the process. During early industrialization, when production sites for large segments of the population shifted from the family and home to the factory, the tradition of family labor and apprenticing children shifted to factories as well. Well-known developments in the countryside during the nineteenth century spurred many families, with their children, to seek employment in cities and industries, a process that further encouraged the shift of child labor to factories. Naturally, widespread cultural acceptance of children's involvement in production and the absence of laws that could have restricted children's employment ensured and stimulated the use of child labor in industry. The crucial role children played in industrialization not only reflected the large number of child workers but also pertained to the actual production process, which, according to entrepreneurs' own testimony, they had designed to function with children's input or which simply operated better with that input. Entrepreneurs were evidently genuinely convinced that their factories could not function properly without children or that without child labor they would suffer economic disadvantage compared to rival industries in countries where child labor was also widespread and unrestricted. At this point, the emphasis fatefully shifted from apprenticeship to actual labor functions, broadening and extending the use of child industrial labor considerably. As a consequence of the harsh realities of child factory labor and in conjunction with the shift from the original pedagogical sociocultural meaning to a harder-edged economic one, the perception of child labor began to shift from social "good" to societal "evil."

Toward the end of the nineteenth century, child labor came under strong and widespread assault. Unlike agriculture or cottage industries, the new industrial setting proved unsafe for children. Not only did children often work in a nonkinship environment, but they were also exposed to the dangers of hazardous machinery and chemicals. Children were also by nature much more susceptible to work-related injuries than adult workers. The obvious precipitous decline in the physical condition of children who worked in industries provoked serious concern and debates about the employment of children in factories, calling into question the notion of apprenticeship as an acceptable practice. Concerned state officials and public figures called for child labor protection laws, although entrepreneurs resisted for economic reasons.

During the late imperial decades, the transformation of attitudes toward the appropriateness of child labor generated considerable changes in the legislation about apprenticeship and children's work. The ongoing public debates during the 1860s and 1870s about children's welfare, employment, and work altered the attitudes of tsarist legislators toward child labor, apprenticeship, and childhood itself. Late tsarist legislators themselves represented a new generation of Russians who were products of a rapidly changing and dynamic society. They were open to inventive legislative ideas and to societal change in general. The attitudes of legislators aside, the ongoing debates among manufacturers, public activists, and legislators provided vital theoretical foundations for the labor protection legislation of the late nineteenth and early twentieth centuries. Unlike the earlier legislation, which had tended to focus on specific factories or other delimited venues, the laws of the 1880s and the following decades became more methodical and wide-ranging. They dealt not only with the minimum workday and employment age but sought to improve children's welfare in general by addressing working conditions and health care and by requiring education for working children. The new regulations also stipulated penalties for employers who transgressed the law (in Russia, as elsewhere, entrepreneurs and business administrators blithely ignored laws without stern penalties).

Russian child labor legislation was of striking significance both as regards the labor question and for society as a whole. The very concept of labor protection legislation in its broadest definition arose in association with discussions of and debates about laws that might regulate children's employment. Only later did the aura of concern extend toward other groups of workers. Here too Russian practice followed a pattern quite common in other countries, whether they introduced labor legislation earlier or later than Russia.[1] This study explores for the first time the laws that limited child labor and introduced compulsory education and welfare; it also strives to place these laws into their proper historical and intellectual context. At the same time, the analysis provided here of the debates about factory-labor laws and the role they played in actual legislation illuminates the process of imperial Russian lawmaking. In fact, it suggests a sharp alteration in how imperial Russian lawmaking should be understood. Contrary to the traditional scholarly view of imperial Russia's strict adherence to hierarchical power relations, the process described here suggests an interactional relationship between the late tsarist state and the developing civil society. The resulting laws reflected society's concerns as expressed in public discourse more than they did autocratic priorities. These laws arose as direct consequences of broad public discussion of child labor and childhood; the laws also reflected compromises that involved state bureaucracies; various political, academic, and business groups; individuals concerned about public welfare; and to some extent even working populations. The process of imperial lawmaking analyzed in this book contributes to a recent tendency in historical studies of late imperial Russia to place ever greater emphasis on society's involvement in legislative processes even before the rise of national elected legislatures.

As mentioned, recent scholarship has begun to challenge the conventional autocracy-centered approach that stresses the power of the state and the tsarist bureaucracy over a weak civil society. Alongside its emphasis on societal involvement and activism, this study suggests novel ways of understanding and interpreting the late imperial Russian state itself. As society became more active and self-assured, the state as often as not responded not by attempting to restrict all impulses from below but by adapting to these impulses in positive ways. Undoubtedly, the late tsarist state was more dynamic, adaptable, and responsive to public pressures than long-established interpretations have allowed. This point does not challenge traditional views about the late tsarist regime's ultimate political failings. It does suggest that we should not exaggerate those failings, as serious as they were. It also suggests a revision of the autocratic concept for Russia, at least during the later imperial era. If one deploys the classic definition, the late tsarist state—its pretensions and its self-proffered public image aside—was hardly or only barely an autocracy.² The authority of Russia's late imperial rulers suffered significantly, in part because of growing institutional and bureaucratic weakness, itself a reflection of a general lack of credibility. Constraints on imperial authority also arose in the form of an assertive and waxing network of interdependent economic, social, and cultural realities that inexorably impinged upon and cut away at autocratic prerogatives. These societal entities functioned as intermediary agencies between the state and society, constraining and influencing in both directions.³

The broad public discussion of the workers' plight and the subsequent passing of legislation aimed at easing that plight suggest a willingness on the part of the tsarist government to rely on measures of amelioration to cope with the labor question, along with the coercive measures that historians habitually emphasize. Arguably, by the end of the period under discussion, these and other laws, including the quite generous and understudied workers' insurance law, had the potential to significantly improve the condition of working families, adults and children. Tragically, the outbreak of World War I defeated all such efforts. Indeed, the harsh economic exigencies of wartime reversed the process by overturning, step by step, the earlier efforts at protecting laboring children and all workers. The hard-won gains of earlier generations of concerned citizens, state officials, and workers themselves in the end went for naught.

Regardless, the rapid urbanization and heavy industrialization that characterized late imperial Russia had a decisive influence on the involvement of children in the social and political events of the nation, including in the early twentieth-century revolutions. As yet, these matters have hardly been broached, much less explored, in existing histories.

Despite the government's efforts to put working children in school, the number of appropriate institutions for children and adolescents always lagged behind the need. Many city children spent their spare time in the streets. Some actively engaged in political activity and even joined radical parties; others manifested their protest of the existing social order through hooliganism.⁴ Working children suffered much of the same exploitation as their adult counterparts and naturally sought redress through various modes of self-expression.

In another set of stunning events, as though Russia and her citizens had not experienced enough trauma, the revolutions and the civil war drastically altered all social equations. Ever more tumultuous military, social, and economic events led to the disintegration of families, as well as the collapse of existing child care and educational systems. Many children entered the tragic state of homelessness (*besprizornost'*). The Communist government's policy of enforced food requisitioning sharply exacerbated the issue, with especially malign results for rural children. At the same time, the collapse of industry led to increased hardship for city children who had formerly been employed in those industries or who were the dependents of factory workers. The resulting wave of children's homelessness reached unheard-of proportions and persisted as one of the great evils of the new regime's first decade. By 1921, according to official government sources, the number of Soviet children either orphaned or hopelessly separated from their families approached six million.[5] But that is another story, perhaps the grimmest one, about childhood in modern Russia.

In his diaries, Dostoevsky reports an incident when, as he strolls through a cityscape, his path intersects for a few moments with that of a young adolescent male of the worker milieu, who with every step rhythmically pronounces a string of Russian curses (*mat*). Whether the famous author perceived this as signifying the degeneration of youth or as a proud accomplishment of Russian folk art, he does not quite say— rather the latter, it would seem. In an act of sympathetic imagination, we might suggest that the young lad, like many of his laboring brothers and sisters, had achieved a sense of independence and self-worth through his work, where he in all likelihood had picked up the curses too. Life lay before him, and he was not afraid. The hard labor of children and youths is not an attractive prospect. Still, many of these young people survived and were even sustained and strengthened. Some sought spiritual sustenance through religion or culture; others became revolutionaries. Still others simply became seasoned workers who married and bore children, who in turn repeated the cycle or, given late imperial Russia's vast social mobility, moved up the socioeconomic scale through education and general urban savvy. Others perished or sank into one or another pathological morass. This study's pages attempt to recapture the experiences of working children and youths in their fullest concrete reality: tragedy and accomplishment in full measure.

A final summary of this study's topic and analysis underscores the reality that the history of late imperial and Soviet Russia is hardly overstudied. Numerous volumes and articles about child labor have appeared within national historiographies of other parts of the world. Studies of aspects of child labor, for example, have appeared about individual states of the United States. The absence until now of any modern study of child labor in Russia is of itself significant, even strange. The lack, however, is not limited to child labor. The empirical record of Russia's history is filled with holes and yawning gaps. As historians begin to fill the empty spaces, their data and analysis routinely challenge existing interpretations in multifarious ways. Any survey of recently published studies indicates how strong this tendency is. None of this is really surprising. British historians are busily recasting interpretations of almost every aspect of their already splendidly researched history. They are doing so on the basis of new archival material and rearguing old cases to drive home new points. Throughout much of the twentieth century, historians of Russia, whether at home or abroad, labored under analytical and source restrictions. By analogy, this study's finding—that close analysis of child labor and pertinent legislation raises questions far beyond the realm of child labor as a socioeconomic reality—hints at further, similar developments. This is already occurring as historians look more closely at phenomena thought to be entirely absent in Russia, or that no one has thought of studying before (like child labor), or that have been studied from within severe restraints on available sources or—almost as bad—from within one or another ideological tendency. The other side of the complaint presents a great opportunity.

Appendix

Documents

1845 Decree on Children's Employment[1]

19262—August 7. Decree approved by his Imperial Highness of the Committee of Ministers and published on September 13.—*On the prohibition of entrepreneurs to employ children under the age of twelve for nighttime work.*

The Governing Senate heard the report of the Minister of Finances, which stated that some businesses conduct work during the day and night and that nighttime work is particularly burdensome for underage workers. In order to alleviate the latter, and following the provisions of the Committee of Ministers, the Sovereign Emperor ... decreed: owners of business that conduct nighttime work are required to sign memorandums that oblige them not to employ children below twelve years of age from midnight to six A.M. The supervision of this law's compliance is to be laid on the local officials. Reporting His Highnesses Will to the Governing Senate, the Minister will give on his behalf orders to implement [His Will].

Decree Approved by His Imperial Highness on the 1st day of June of 1882—*On the opinion of the State Council about the measures of restriction of work of children and juveniles in factories, works, and other industrial units and about their education.*[2]

I. In the change of and addition to the appropriate articles of the Law Code about children of both sexes who work in factories, plants, and manufacturing establishments that belong to private individuals and organizations (societies, associations, and companies), as well as to the state, the following rules are introduced:

1. Children under the age of twelve are not allowed for employment.

2. Children between the ages of twelve and fifteen cannot work more than eight hours a day, not including time for breakfast, lunch, and dinner; attendance at school; and rest. Their work cannot last more than four consecutive hours.

3. Children under the age of fifteen cannot work between nine P.M. and five A.M., as well as on Sundays and major holidays.

4. The children mentioned in Article 3 are prohibited from employment in such industries or for single works that are parts of these industries that are by their nature harmful or recognized as exhausting for health. The list of such industries and occupations . . . is to be defined by the mutual agreement of the Finance and Interior ministries. . . .

5. The owners of factories, plants, and manufacturing establishments are required to provide their working children who have no diploma at least the one-class program of public or its equivalent school with no less than three hours a day or eighteen hours a week in order to attend said schools.

II. In order to control the implementation of the regulations of labor and education of child workers, a special inspection is introduced on the following grounds:

1. Regarding the control over work and education of child workers, the areas with industries are to be divided into special districts. Their number, as well as the arrangement of provinces and areas within each district, is to be approved by law.

2. Depending on the necessity, each district has one or several inspectors. The general supervision over all districts is handed over to the chief inspector. This inspectorate is placed under the authority of the Ministry of Finance's Trade and Manufacture Department.

3. The district inspectors are responsible to the chief inspector and relate to the provincial and local authorities on the same basis as all other officials of the Finance Ministry who belong to the provincial government. . . .

4. Inspectors are obliged to (1) oversee compliance with the laws on child labor and the education of working children; (2) file with the local police protocols about violations of the said laws and submit these protocols to the appropriate legal institutions; and (3) bring to court persons responsible for violations. . . .

5. Detailed provisions for responsibilities and procedures are set up in a special instruction to be approved by the Ministry of Finance with the agreement of the ministries of the Interior and Education.

6. The authority of the inspectorate ... does not extend to factories, plants, and manufacturing establishments that belong to the state or government. Control over labor and education of children employed in these enterprises is given to those appointed persons who manage them.

III. The provisions of Part One are to be enacted on May 1, 1883.

Accidents of Working Children Recorded by the Local District Police Offices in Moscow

Accidents at the Wool Cloth Factory of Merchant Nosov in the Spring of 1857

Testimony of worker Petr Afanas'ev[3]

On April the 2nd of 1857 in Lefortovo district police house the following were questioned and testified:

I am Petr Afanas'ev, eighteen years old. By belief I am Orthodox [Christian]. I go to confession at a sacred participle every year and know literacy. I am the peasant of landlord Durnovo of the village Klemova in the Venevsk district of the Tula province. I was not under investigation. I am among the factory workers at the merchant Nosov's mill. My work consists of supervising the boys who stand at shearing machines and at each machine arrange the cloth going onto the shaft. These are up to six such boys. Last March, the 23rd, at eight o'clock in the morning, right before breakfast, one of the boys, the peasant Andrei Agapov, somehow, I do not know for sure how, got his hand into the machine and had two middle fingers injured. As I dare say, this may have been because of his own imprudence, for I strictly observe all of them and nobody is engaged in pranks. Everyone stands at the machines in their places. . . .

To this deposition the peasant Peter Afanas´ev affixed his hand.

District police officer (signature)

Testimony of the boy Andrei Agapov[4]

On April the 5th, 1857, in Lefortovo, the district police house sent by the hospital for workers peasant boy Andrei Agapov, sixteen years old, of landlord Vasil'chikov, who testified during the questioning:

By faith I am Orthodox Christian [and] take Holy Communion every year. I know literacy, but because of the disorder of my right hand, on which the fingers were injured, I cannot affix my signature. I am living at the mill of the merchant Nosov since the autumn of last 1856 [and work] as a helping boy on the shearing machine. Last March, the 23rd, right before breakfast, when I was on duty with my fellow worker Nikifor Nikiforov, I tried to straighten the cloth when it began to jam.... Two fingers of my right hand went with the cloth on the knives that cut nap. These knives cut off the nail to the bone on my middle finger and cut off the flesh to the bone on the fourth one. I had pulled out my hand and was so scared that I did not feel any pain until the local physician arrived and dressed the wounds. I was immediately sent to a hospital.... Meanwhile I am feeling all right and ask to be sent back home from where my father has arrived to take me. This was an accident, and I do not blame anyone else for doing this deliberately.

Due to the hand injury, this testimony is notarized by Moscow *meshchanin* Mikhaila Ignatiev Shurov according to Agapov's personal favor.

District police officer (signature)

The receipt of peasant Agap Andreev[5]

On April, 5th, 1857, I, the undersigned peasant of the village Shebanovo of the District Aleksinsky, Tula province, of landowner Vasilchikov Agap Andreev, have given this receipt that I have accepted my son Andrey Agapov who was at the factory of Nosov.... On the merchant Nosov I have no claims beyond the reception of satisfaction. I know literacy.

The peasant of landlord Vasilchikov Agap Andreev has put his hand to this receipt.

District police officer (signature)

Notes

Introduction

1. Nekrasov, *Sobranie sochinenii v vos'mi tomakh*, 1: 359.
2. Child labor is described in A. Chekhov's story "Spat' Khochetsia"; in Maxim Gorky's novels *Mat'*, *V liudiakh*, and other works; in E. Nechiaev's poem "Gudok"; in the novels of A. Kuprin (*V nedrakh zemli*) and L. Serafimovich (*Pod prazdnik*); and in the works of many other poets and writers of the time.
3. Cited in Pazhitnov, *Polozhenie*, 2: 28.
4. For a discussion of child labor in general, see Hobbs et al., *Child Labor*. On child labor in England, see Horn, *Victorian and Edwardian Schoolchild*; Nardinelli, *Child Labor*; Horn et al., *Children's Work and Welfare*; Tuttle, *Hard at Work*; Kirby, *Child Labor*. Child labor in France is discussed in Heywood, *Childhood in Nineteenth-Century France*; and Weissbach, *Child Labor Reform*. For a bibliography on child labor in the United States, see Hindman, *Child Labor*. On child labor in the developing world, see Rodgers and Standing, *Child Work, Poverty and Underdevelopment*; Post, *Children's Work, Schooling and Welfare*.
5. For examples of this interpretation, see Fuller, "Child Labor"; Thompson, *Making of the English Working Class*; Hammond and Hammond, *Town Labourer*.
6. Walvin, *Child's World*, 64.
7. Thompson, *Making of the English Working Class*, 349.
8. For these views, see Hartwell, *Industrial Revolution and Economic Growth*.
9. For further discussion, see Nardinelli, *Child Labor*, 98 (quotation on 102).
10. See Horrell and Humphries, "'Exploitation of Little Children.'"
11. Andreev, *Rabota*. Other late imperial historians also explore various aspects of child factory labor. Tugan-Baranovsky, e.g., discusses the 1845 child labor law in *Russian Factory*. For the Russian-language edition, see Tugan-Baranovsky, *Russkaia fabrika v proshkom i nastoiashchem*. E. M. Dement'ev also discusses some aspects of child labor in his *Fabrika*.
12. Gessen, *Trud*; Gessen, *Istoriia zakonodatel'stva*. Child labor in imperial Russia is also discussed in Zaitsev, *Polozhenie*; Kirov, *Na zare*. The issue of children's employment in industries is also mentioned in many Russian-language studies on workers and the labor movement. In this literature, however, the issue did not receive any specific analysis. The only recent Russian-language account of child labor is an essay by N. I. Enalieva,

whose interest is education. This seven-page general overview, whose usefulness is limited by a lack of references, has been translated into English. See Enalieva, "Child Labor and Industrial Apprenticeship in Russia."

13. See Zelnik, *Labor and Society*; Zelnik, *Law and Disorder on the Narova River*, 26, 72, 132, 169, 174, 228–29.

14. Melancon, *Lena Goldfields Massacre*.

15. Kelly, *Children's World*, 3.

16. Here is where my findings modify Kelly's suggestion about the beginnings of child-related discourse in public, professional, and state institutions.

17. Marxist-inspired scholarship that underscores class includes, among others, Reichman, *Railway Men and Revolution*; Surh, *1905 in St. Petersburg*; Zelnik, *Labor and Society*; Johnson, *Peasant and Proletarian*. For an overview of Russian-language scholarship, see Pankratova, *Formirovanie proletariata v Rossii*; Rashin, *Formirovanie*.

18. For this scholarship, see, among others, Glickman, *Russian Factory Women*; Engel, *Between the Fields and the City*.

Chapter 1

1. Gmelin, *Reise durch Sibirien*, cited in Pazhitnov, *Polozhenie*, 1: 56.

2. Pallas, *Puteshestviia po raznym provintsiiam Rossiiskago gosudarstva*, cited in Pazhitnov, *Polozhenie*, 1: 56.

3. For a discussion of the origins of child labor in France, see Weissbach, *Child Labor Reform*, chap. 1; Heywood, *Childhood*, 34.

4. Gromyko, *Mir russkoi derevni*, 106.

5. For example, Colin Heywood emphasizes the poor conditions of peasant families in France as a principal source of child labor in the French countryside. See Heywood, *Childhood in Nineteenth-Century France*, 38.

6. Gromyko, *Trudovye traditsii russkikh krest'ian Sibiri*, 4–5; Minenko, *Russkaia*, 117.

7. For further examples of children's collective names, see Bernshtam, *Molodezh*, 25, 122.

8. Aries, *Centuries of Childhood*, 47.

9. For discussion of this issue, see Bernshtam, *Molodezh*, 24–25.

10. Shangina, *Russkie deti i ikh igry*, 7.

11. Baklanova, *Krest'ianskii*, 23.

12. Baklanova, *Krest'ianskii*, 23, 41–42; Bernshtam, *Molodezh*, 123; Aleksandrov, *Sel'skaia obshchina v Rossii*, 206–7; Lenin, *Razvitie kapitalizma v Rossii*, 3: 325.

13. Baklanova, *Krest'ianskii*, 22 (table 6).

14. Mironov, *Sotsial'naia istoriia Rossii*, 1: 20, 129, 180.

15. Mironov, *Sotsial'naia istoriia Rossii*, 199–200; Baklanova, *Krest'ianskii*, 23.

16. Minge-Kalman, "Industrial Revolution"; Bernshtam, *Molodezh*, 57–58; Gromyko, *Mir russkoi derevni*, 107; Minenko, *Russkaia*, 51–52.

17. Gorshkov, *Life under Russian Serfdom*.

18. Martynova, "Otrazhenie deistvitel'nosti v krest'ianskoi kolybel'noi pesne," 152; Romanov, *Liudi i nravy Drevnei Rusi*, 155.

19. This data is based on the 1717 census of 1,064 families of serfs and monastery peasants of the Kubenskii region of Vologda province, northern Russia, cited in Baklanova, *Krest'ianskii*, 22.

20. Shangina, *Russkie deti i ikh igry*, 7.
21. Bernshtam, *Molodezh*, 25.
22. Bernshtam, *Molodezh*, 57.
23. Bernshtam, *Molodezh*, 56.
24. Bernshtam, *Molodezh*, 55.
25. Bernshtam, *Molodezh*, 56.
26. Bernshtam, *Molodezh*, 53.
27. Bernshtam, *Molodezh*, 58.
28. For further discussion of children's clothing, see Maslova, *Narodnaia odezhda*, 106.
29. Aries, *Centuries of Childhood*.
30. Gromyko, *Mir russkoi derevni*, 108.
31. For further discussion of rural children's games, see Shangina, *Russkie deti i ikh igry*.
32. Cited in Minenko, *Russkaia*, 118.
33. Bernshtam, *Molodezh*, 122–23.
34. Minenko, *Russkaia*, 118.
35. Minenko, *Russkaia*, 118.
36. Bernshtam, *Molodezh*, 25, 122.
37. For a discussion of peasant seasonal migration before 1861, see Gorshkov, "Serfs on the Move."
38. Minenko, *Russkaia*, 120.
39. This area, known as the Central Industrial Region, included the provinces of Yaroslavl', Tver', Kostroma, Kaluga, Moscow, Vladimir, Nizhnii Novgorod, and usually Tula and Riazan'.
40. Gorshkov, *Life under Russian Serfdom*, 64.
41. Bernshtam, *Molodezh*, 123.
42. Chebotarev, *Istoricheskoe i topographicheskoe opisanie gorodov*, 119–348.
43. Gromyko, *Traditsionnye*, 109.
44. Bernshtam, *Molodezh*, 123.
45. Cited in Minge-Kalman, "Industrial Revolution," 455.
46. Minge-Kalman, "Industrial Revolution," 456.
78. Cited in Kirby, *Child Labor*, 15.
48. See discussion in Nardinelli, *Child Labor*, 51–57.
49. For further discussion of these enterprises, see Pazhitnov, *Polozhenie*, 1: 16–52; Tugan-Baranovsky, *Russian Factory*, 82–131; Blum, *Lord and Peasant*, 308–25; Isaev, *Rol' tekstil'noi promyshlennosti*, 118–69; Kazantsev, *Rabochie Moskvy*, 75–76; Gorshkov, "Serfs on the Move."
50. Cited in Gessen, *Istoriia zakonodatel'stva*, 55. For biographic information about N. S. Mordvinov and A. A. Zakrevskii, see Shilov, *Gosudarstvennye*, 246–49, 432–37. For an informative discussion of N. S. Mordvinov and his ideas, see McCaffray, "What Should Russia Be?"
51. Tugan-Baranovsky, *Russian Factory*, 138.
52. *Polnoe sobranie zakonov Rossiiskoi Imperii*, 1st ser. (hereafter *PSZ* 1), vol. 35 (1818), no. 27438.
53. "Donesenie upravliaiushchego peterburgskim liteinym zavodom," in Trofimov, *Rabochee*, 205.
54. *PSZ* 1, vol. 29, 1806, no. 22099. See also Gessen, *Trud*, 32.
55. Gessen, *Trud*, 39.

56. Kirov, *Na zare*, 9–10.
57. *PSZ* 1, no. 21368.
58. Balabanov, *Ocherki*, 65; Gessen, *Istoriia zakonodatel'stva*, 14.
59. Gessen, *Trud*, 29.
60. Gessen, *Istoriia zakonodatel'stva*, 14; *PSZ* 1, vol. 6, no. 4006.
61. Kirov, *Na zare*, 9.
62. Gessen, *Istoriia zakonodatel'stva*, 29.
63. Kitanina, *Rabochie*, 170.
64. Gessen, *Trud*, 34.
65. Cited in Pazhitnov, *Polozhenie*, 57; Kirov, *Na zare*, 10.
66. Gessen, *Trud*, 22.
67. *PSZ* 1, nos. 22099, 27438.
68. For example, state decrees proclaimed that children of factory workers were supposed to take employment according to their "age, gender, and strength" (*PSZ* 1, nos. 22099, 27438); Gessen, *Trud*, 25.
69. Pazhitnov, *Polozhenie*, 1: 56.
70. Tsentral'nyi gosudarstvennyi istoricheskii arkhiv Moskvy (hereafter TsIAM), *fond* 16, *opis'* 34, *delo* 48, *listy* 1–25; Gessen, *Trud*, 35, 196–97.
71. *PSZ* 1, no. 26416; Tugan-Baranovsky, *Russkaia fabrika*, 95.
72. Gessen, *Trud*, 33.
73. Gessen, *Trud*, 40–41.
74. *PSZ* 1, no 18965; Pazhitnov, *Polozhenie*, 1: 56.
75. *PSZ* 1, nos. 22099, 27438.
76. *PSZ* 1, no. 21368.
77. All male peasants were obliged to pay the state poll tax. Children from seven to seventeen paid from 0.15 to 1.60 rubles a year, depending on their age (Gessen, *Trud*, 26).
78. "Prosheniia rabochikh bol'shoi Yaroslavskoi Manufactury," in Trofimov, *Rabochee*, 161.
79. Pazhitnov, *Polozhenie*, 1: 56–57.
80. Kitanina, *Rabochie*, 170.
81. The data is cited in Bukhina, "Maloiaroslavetskaia fabrika."
82. Gessen, *Istoriia zakonodatel'stva*, 28; Gessen, *Trud*, 33.
83. Kitanina, *Rabochie*, 169.
84. Gessen, *Trud*, 38; Kirov, *Na zare*, 11.
85. Gessen *Trud*, 39; Planer, *Istorikoisticheskoe*, 12.
86. Gessen, *Trud*, 33.
87. According to Pazhitnov, only 391,000 out of the 195,000 hereditary serfs of the Altai region mines and metallurgical plants recorded in the tenth census (1857) actually worked in the enterprises (*Polozhenie*, 1: 34, 60). The observation about the labor of children in state industries also finds support in Kirov, whose 1924 study of child labor suggests that children under age ten were rarely employed in manorial and state factories (*Na zare*, 12).
88. Pazhitnov, *Polozhenie*, 1: 84.
89. Gessen, *Trud*, 121.
90. A. M. Pankratova, *Formirovanie proletariata v Rossii*, 24.
91. Gessen, *Istoriia zakonodatel'stva*, 42; Gessen, *Trud*, 120.
92. TsIAM, *fond* 16, *delo* 48, *opis'* 34, *listy* 1–25.

93. TsIAM, *fond* 16, *delo* 48, *opis'* 34, *listy* 1–25.
94. TsIAM, *fond* 16, *delo* 48, *opis'* 34, *listy* 1–25; Pazhitnov, *Polozhenie*, 1: 59; Gessen, *Trud*, 35, 198; Gessen, *Istoriia zakonodatel'stva*, 50.
95. Gessen, *Trud*, 38–39.
96. Pazhitnov, *Polozhenie*, 1: 56.
97. Brandenburgskii, "Zheleznye zavody."
98. Gessen, *Trud*, 26.
99. Gessen, *Trud*, 38.
100. Gosudarstvennyi arkhiv Rossiiskoi Federatsii (hereafter GARF), *fond* 102, *opis'* 42, *delo* 34 (2), *listy* 25–30. For discussion of the 1835 law, see Gorshkov, "Toward a Comprehensive Law," 54.
101. For discussion of this issue, see Gorshkov, "Serfs on the Move."
102. Gessen, *Istoriia zakonodatel'stva*, 38.
103. GARF, *fond* 102, *opis'* 42, *delo* 34 (2), *listy* 25–26; Gessen, *Trud*, 168.
104. Gessen, *Trud*, 34.
105. Tugan-Baranovsky, *Russkaia fabrika*, 173; Gessen, *Trud*, 64.
106. Laverychev, *Tsarism*, 16.
107. Tugan-Baranovsky, *Russian Factory*, 141.
108. Cited in Bukhina, "Iz istorii," 117–18.
109. TsIAM, *fond* 16, *delo* 48, *opis'* 34, *listy* 1–25; Gessen, *Istoriia zakonodatel'stva*, 50.
110. TsIAM, *fond* 16, *delo* 48, *opis'* 34, *listy* 1–25; Tugan-Baranovsky, *Russkaia fabrika*, 175; Pazhitnov, *Polozhenie*, 1: 59; Gessen, *Istoriia zakonodatel'stva*, 50.
111. *PSZ*, 2nd ser. (hereafter *PSZ 2*), no 19262; Pazhitnov, *Polozhenie*, 1: 60.
112. Tugan-Baranovsky, *Russkaia fabrika*, 139.
113. Kirov, *Na zare*, 12.
114. Gessen, *Istoriia zakonodatel'stva*, 64–65.
115. Pazhitnov, *Polozhenie*, 1: 29.
116. For discussion of early labor laws in Europe, see Hepple, *Making of Labor Law*, chap. 2; Weissbach, *Child Labor Reform*, 123.
117. Tugan-Baranovsky, *Russian Factory*, 144.
118. TsIAM, *fond* 17, *opis'* 34, *delo* 48, *list* 244.

Chapter 2

1. For data and sources regarding the rise in contracted labor as opposed to manorial or attached labor, see Gorshkov, "Serfs on the Move."
2. On Russian economic development during the nineteenth century, see Blackwell, *Beginnings*; Crisp, *Studies*; Kahan, *Russian Economic History*; Liashchenko, *History of the National Economy*; Tugan-Baranovsky, *Russian Factory*, 48.
3. Early Russian industrialization receives treatment in Blackwell, *Beginnings*; Crisp, *Studies*; Kahan, *Russian Economic History*; Tugan-Baranovsky, *Russian Factory*. For discussion of the textile industry, see Khromov, *Ocherki ekonomiki*; Isaev, *Rol' tekstil'noi promyshlennosti*.
4. For discussion of the labor force before 1861, see Rashin, *Formirovanie*. See also Gorshkov, "Serfs on the Move," 635 n32, 639–41. Some labor issues before 1861 are addressed in Zelnik, *Labor and Society*.
5. Kahan, *Russian Economic History*, 48.

6. Kahan, *Russian Economic History*, 3. See also Mironov, *Sotsial'naia*, 1: 315.

7. On the issue of labor migration from the countryside during the late imperial period, see Johnson, *Peasant and Proletarian;* Brower, *Russian City;* Engel, *Between the Fields and the City;* Economakis, "Patterns of Migration." See also Burds, *Peasant Dreams,* among other studies.

8. The 1897 census data is cited in Kabanov, Erman et al., *Ocherki*, 21. According to A. G. Rashin, in 1897, about 12 percent of industrial workers were children under fifteen years of age (*Formirovanie*, 280).

9. The figure for industrial workers includes workers in factories, mines, and railroads. Cited in Kabanov, Erman et al., *Ocherki*, 21, 23.

10. Rashin, *Formirovanie*, 256.

11. Gorshkov, "Factory Children," 17.

12. TsIAM, *fond* 17, *opis'* 34, *delo* 48, *list* 244. See also Gorshkov, "Serfs on the Move," 644.

13. Gessen, *Trud*, 46.

14. *Proekt pravil*, 197–200. See also Kitanina, *Rabochie*, 172.

15. Gessen, *Trud*, 46.

16. "Vnutrennee obozrenie," *Vestnik Evropy* 5, no. 10 (Oct. 1875): 801–26, 824.

17. Andreev, *Rabota*, 43–49.

18. TsIAM, *fond* 1780, *opis'* 1, *delo* 14, *listy* 4, 110–12.

19. TsIAM, *fond* 1780, *opis'* 1, *delo* 3; Rashin, *Formirovanie*, 251.

20. Rashin, *Formirovanie*, 252.

21. Gessen, *Trud*, 46.

22. Rashin, *Formirovanie*, 256.

23. Andreev, *Rabota*, 1–160, appendix.

24. Gessen, *Trud*, 56.

25. Peskov, *Fabrichnyi byt*, 6.

26. Anokhina and Shmeleva, *Byt gorodskogo*, 63.

27. Peskov, *Fabrichnyi byt*, 16; Mikhailovskii, *O deiatel'nostii*, 14.

28. Leverychev, *Tsariam*, 3.

29. Rashin, *Formirovanie*, 251.

30. Gessen, *Trud*, 47.

31. Rashin, *Formirovanie*, 251.

32. Rashin, *Formirovanie*, 251.

33. Gessen, *Trud*, 48.

34. Zaitsev, *Polozhenie*, 32.

35. Cited in Rashin, *Formirovanie*, 251.

36. Peskov, *Fabrichnyi byt*, 24.

37. Gessen, *Trud*, 50.

38. Gessen, *Trud*, 46–49.

39. For discussion of the Russian textile industry, see Khromov, *Ocherki ekonomiki*.

40. Cited in Andreev, *Rabota*, 14.

41. Cited in Rashin, *Formirovanie*, 251.

42. Gessen, *Trud*, 49.

43. Kirov, *Na zare*, 40.

44. Peskov, *Fabrichnyi byt*, 39.

45. The cases of these families are cited in Andreev, *Rabota*, 167–73.

46. Zaitsev, *Polozhenie*, 29.
47. Andreev, *Rabota*, 172.
48. Andreev, *Rabota*, 168–69.
49. Cited in Zaitsev, *Polozhenie*, 29.
50. Cited in Rashin, *Formirovanie*, 260.
51. Milogolova, "Semeinye," 37. For an English-language study of this issue, see Frierson, "Razdel."
52. Cited in Milogolova, "Semeinye," 37.
53. Tiukavkin, *Velikorusskoe*, 35; see table 1.
54. Kabuzan, *Russkie v mire*; Tiukavkin, *Velikorusskoe*, 34.
55. See, e.g., Kirby, *Child Labor*, 30.
56. Milov, "Otkhodnichestvo"; cited in Tiukavkin, *Velikorusskoe*, 54.
57. For further discussion of these workers' associations, see Johnson, *Peasant and Proletarian*; Zelnik, *Labor and Society*; Gorshkov, "Serfs on the Move," 645–46; Worobec, *Peasant Russia*.
58. Peskov, *Fabrichnyi byt*, iii; Mikhailovskii, *O deiatel'nosti*, 16–17.
59. Tiukavkin, *Velikorusskoe*, 55; Lenin, *Polnoe sobranie sochinenii*, 3: 518–25.
60. Gessen, *Trud*, 49; Dement'ev, *Fabrika*, 147–48.
61. Andreev, *Rabota*, 38; Gessen, *Trud*, 50.
62. Andreev, *Rabota*, 181.
63. Cited in Kirov, *Na zare*, 33–34.
64. Dement'ev, *Fabrika*, 147–48.
65. Peskov, *Fabrichnyi byt*, 38–43.
66. Dement'ev, *Fabrika*, 46.
67. Nardinelli, *Child Labor*, 106, table 5.2.
68. Weissbach, *Child Labor Reform*, 165, table 4.
69. Salle, *Whiteness*, 31.
70. See Peskov, *Fabrichnyi byt*; see *prilozheniia*, 3–53, for the Sokolovskaia mill.
71. Peskov, *Fabrichnyi byt*; *prilozheniia*, 3–53.
72. Peskov, *Fabrichnyi byt*, 8, 44.
73. Peskov, *Fabrichnyi byt*, 46.
74. Peskov, *Fabrichnyi byt*, 47.
75. Zaitsev, *Polozhenie*, 32–33.
76. Andreev, *Rabota*, 181.
77. Peskov, *Fabrichnyi byt*, 46–47. These figures on working hours are supported by the data gathered by the Commission for Technical Education of the Imperial Russian Technical Society in 1875, as presented in Andreev, *Rabota*, 46.
78. Gessen, *Trud*, 48.
79. Zaitsev, *Polozhenie*, 17.
80. Cited in Andreev, *Rabota*, 14.
81. Andreev, *Rabota*, appendix, 4.
82. Zaitsev, *Polozhenie*, 18.
83. Andreev, *Rabota*, 168–73.
84. Cited in Gessen, *Trud*, 219.
85. Peskov, *Fabrichnyi byt*, 42.
86. Blakhin, "Zhizn' fabrichnogo rabochego," in *Biblioteka dlia chtenia*, 71: 63; Kitanina, *Rabochie*, 235; Kir'ianov, *Zhiznennyi*, 70.

87. Peskov, *Fabrichnyi byt*, 107.
88. Merriman, *From the French Revolution*, 870.
89. Peskov, *Fabrichnyi byt*, 104; Andreev, *Rabota*, 81; Gessen, *Trud*, 46.
90. Dement'ev, *Fabrika*, 238.
91. Andreev, *Rabota*, 181 (in *prilozhennia*).
92. Andreev, *Rabota*, 182–83; Bukhina, "Iz istorii," 84.
93. Peskov, *Fabrichnyi byt*, 84.
94. Peskov, *Fabrichnyi byt*, 84.
95. Peskov, *Fabrichnyi byt*, 84.
96. Peskov, *Fabrichnyi byt*, 44.
97. Bukhina, "Iz istorii," 129, 131.
98. Bukhina, "Iz istorii," 137.
99. Nardinelli, *Child Labor*, 9–34.
100. Dement'ev, *Fabrika*, 248–49.
101. Chekhov, *Izbrannye*, 1: 139.
102. Zaitsev, *Polozhenie*, 18; Balabanov, *Ocherki*, 126.
103. Gessen, *Istoriia zekonodatel'stva*, 75.
104. Peskov, *Fabrichnyi byt*, 107.
105. N. N. Voskoboinikov, "Neskol'ko voprosov o polozhenii fabrichnykh rabochikh preimushchestvenno na peterburgskikh fabrikakh," in *Biblioteka dlia chteniia*, 171: 233.
106. Bukhina, "Iz istorii," 139.
107. Bukhina, "Iz istorii," 135.
108. Bukhina, "Iz istorii," 141.
109. Peskov, *Fabrichnyi byt*, 125–36.
110. Kitanina, *Rabochie*, 230–31.
111. Cited in Andreev, *Rabota*, 12.
112. Bukhina, "Iz istorii,"132.
113. Bukhina, "Iz istorii," 133.
114. Bukhina, "Iz istorii," 128–30.
115. Bukhina, "Iz istorii," 133.
116. Peskov, *Fabrichnyi byt*, 125–36; see also *prilozheniia*, table 6, pp. 57–67.
117. *Proekt pravil*, 125; Kitanina, *Rabochie*, 232.
118. TsIAM, *fond* 1780, *opis'* 1, *delo* 3, *list* 1a.
119. GARF, *fond* 102, *opis'* 42, *delo* 34 (1), *list* 76.
120. Kitanina, *Rabochie*, 233.
121. Kitanina, *Rabochie*, 233.
122. Kowler and Martins, "Eye Movements."
123. Sugarman, *Life-Span Development*; Hetherington, et al., *Child Development*; Paterson, Heim et al., "Development of Structure."
124. "Vnutrennee obozrenie," 826, 824.
125. Gessen, *Trud*, 48.
126. TsIAM, *fond* 1780, *opis'* 1, *delo* 3, *listy* 1a–2.
127. Cited in Nikol'skii, "K voprosy," 630–31.

Chapter 3

1. GARF, *fond* 102, *opis'* 42, *delo* 34 (1), *list* 74.
2. Zelnik, *Labor and Society*, 119–59; Blackwell, *Beginnings*, 345–51; Laverychev, *Tsarism*, 14–15.
3. GARF, *fond* 102, *opis'* 42, *delo* 34 (1) *list* 76.
4. GARF, *fond* 102, *opis'* 42, *delo* 34 (1) *listy* 25–26.
5. Rosiiskii gosudarstvennyi istoricheskii arkhiv (hereafter RGIA), *fond* 504, *opis'* 1, *delo* 99, *listy* 5–20; Andreev, *Rabota*, 5, 12.
6. Cited in Andreev, *Rabota*, 13.
7. GARF, *fond* 102, *opis'* 42, *delo* 34 (1), *list* 76.
8. GARF, *fond* 102, *opis'* 42, *delo* 34 (1), *list* 76; Andreev, *Rabota*, 4–5. In his studies on child labor and related legislation, the Soviet historian Gessen stated that the proposals limited the employment age to ten and the workday for children ages ten to twelve to six hours and for children between twelve and fourteen to twelve hours. Other sources, including archival documents and those cited in Andreev, do not support Gessen's statement. The twelve-hour workday suggested by the commissions included a two-hour break for lunch and rest so that actual working hours were limited to ten. See Gessen, *Istoriia zakonodatel'stva*, 56–57; Gessen, *Trud*, 52–53.
9. Andreev, *Rabota*, 5, 12.
10. GARF, *fond* 102, *opis'* 42, *delo* 34 (1), *list* 76; Andreev, *Rabota*, 5.
11. GARF, *fond* 102, *opis'* 42, *delo* 34 (1), *list* 77.
12. Gessen, *Istoriia zakonodatel'stva*, 58; Laverychev, "Iz istorii," 66. There were basic differences between *arteli* and *zemliachestva*. *Arteli* were often small groups of peasant-migrants who sought temporary employment. Some of these *arteli* included only children. Workers' *zemliachestva* were based on regional identities and were created in urban areas at large enterprises; they were usually stationary and large in membership. For discussion of these associations and peasant migrants, see Gorshkov, "Serfs on the Move," 645. See also Rudolph, "Agricultural Structure"; Moon, "Peasant Migration."
13. Ivanov and Volin, *Istoriia rabochego*, 123.
14. GARF, *fond* 102, *opis'* 42, *delo* 34 (1), *listy* 76–77.
15. Nardinelli, *Child Labor*, 104.
16. For further discussion of the early factory laws, see chapter 1. Laura Engelstein notes that "all ... lawmakers in Russia and the West borrowed from each other's laws ("Combined Underdevelopment," n27).
17. GARF, *fond* 102, *opis'* 42, *delo* 34 (1), *list* 77.
18. Vladimir Gessen, e.g., argued that the St. Petersburg commission's legislative proposal was "defeated because of the firm opposition of Muscovite entrepreneurs" (*Trud*, 53).
19. In his *Tsarism i rabochii vopros*, Laverychev points out that St. Petersburg industrialists found the minimum age for employment "not quite desirable" (19).
20. GARF, *fond* 102, *opis'* 42, *delo* 34(1), *list* 76.
21. Ivanov and Volin, *Istoriia rabochego*, 123–24.
22. Pazhitnov, *Polozhenie*, 2: 23; Gessen, *Trud*, 52.
23. Andreev, *Rabota*, 15–16; Gessen, *Istoriia zelonodatel'stva*, 56–57.
24. Andreev, *Rabota*, 15–16; Gessen, *Istoriia zelonodatel'stva*, 56–57.
25. Gessen, *Istoriia zakonodatel'stva*, 57.

26. Gessen, *Istoriia zakonodatel'stva*, 57.
27. Gessen, *Trud*, 47; Gessen, *Istoriia zakonodatel'stva*, 89.
28. Gessen, *Istoriia zakonodatel'stva*, 56.
29. Quoted in Laverychev, "Iz istorii," 66.
30. Laverychev, "Iz istorii," 67.
31. *Trudy komissii*, 274–78, Andreev, *Rabota*, 5–6, 12, 16.
32. Gessen, *Istoriia zakonodatel'stva*, 57.
33. *Trudy komissii*, 274–78.
34. Cited in Gessen, *Istoriia zekonodatel'stva*, 57.
35. *Trudy komissii*, 274–75.
36. Laverychev, *Tsarism*, 19.
37. Nardinelli, *Child Labor*, 133.
38. Nardinelli, *Child Labor*, 275.
39. Nardinelli, *Child Labor*, 278.
40. *Trudy komissii*, 287; Gessen, *Istoriia zakonodatel'stva*, 58.
41. TsIAM, *fond* 2354, *opis'* 1, *delo* 49, *list* 38; Gessen, *Istoriia zakonodatel'stva*, 57.
42. Andreev, *Rabota*, 15–16; Gessen, *Trud*, 47.
43. TsIAM, *fond* 2354, *opis'* 1, *delo* 49, *list* 38.
44. Andreev, *Rabota*, 14.
45. Gorshkov, "Toward a Comprehensive Law," 58. In this article, I mistakenly claimed that hospital beds were to be established at the rate of five beds for every one thousand employees. In reality, hospital beds were set up at the rate of one for every one hundred employees. See Zhuk, *Razvitie*, 327–28; Ivanov and Volin, *Istoriia rabochego*, 124.
46. Laverychev, "Iz istorii," 24; Laverychev, *Tsarism*, 25–26.
47. Cited in Laverychev, "Iz istorii," 65.
48. Tugan-Baranovsky, *Russkaia fabrika*.
49. TsIAM, *fond* 143, *opis'* 1, *delo* 34, *list* 4; TsIAM, *fond* 2354, *opis'* 1, *delo* 49, *list* 38.
50. See, e.g., Rieber, *Merchants*, 421; Owen, *Capitalism*, 421.
51. German sociologist Juergen Habermas suggested the concept of an "informal voluntary public sphere in civil society where private middle class individuals joined in groups became involved in 'critical' and 'reasoned' discourse about common issues . . . in order to influence the process of state decision making." For Habermas, this represented one of the aspects of democratic government (cited in Gorshkov, "Democratizing Habermas," 374). For further discussion of the public sphere, see Habermas, *Structural Transformation*; Habermas, *Philosophical Discourse*.
52. GARF, *fond* 102, *opis'* 42, *delo* 34 (1), *list* 111.
53. Kabanov, Erman et al., *Ocherki*, 37–38.
54. Laverychev, *Tsarism*, 14–15, Zelnik, *Labor and Society*.
55. GARF, *fond* 102, *opis'* 42, *delo* 34 (1), *list* 78.
56. RGIA, *fond* 1149, *opis'* 9, *delo* 31 (1), *listy* 53–54; GARF, *fond* 102, *opis'* 42, *delo* 34 (1), *list* 77; Gessen, *Trud*, 48.
57. RGIA, *fond* 1149, *opis'* 9, *delo* 31(1), *listy* 3–26; GARF, *fond* 102, *opis'* 42, *delo* 34 (1), *list* 78.
58. P. Paradizov, "'Rabochii vopros' v Rossii v nachale 70-kh godov XIX v.," in Pankratova, *Istoriia proletatiata*, 9: 55.
59. Paradizov, "'Rabochii vopros,'" 62.
60. Paradizov, "'Rabochii vopros,'" 62.

61. *Protokoly i stengraficheskie otchety;* Andreev, *Rabota,* 41–42; Gessen, *Istoriia zakonodatel'stva,* 70.

62. Paradizov, "'Rabochii vopros'," 55.

63. Cited in Gessen, *Istoriia zakonodatel'stva,* 71.

64. The idea that child labor led to the decline of wages for adult workers was advanced by Karl Marx. For discussion, see *Kapital,* 1: 395–402.

65. Paradizov, "'Rabochii vopros'," 61.

66. Kaigorodov was one of the early activists of Russia's labor movement. For further discussion, see Paradizov, "'Rabochii vopros'," 59.

67. Pankratova, *Istoriia proletariata,* 60.

68. Gessen, *Istoriia zaonkodatel'stva,* 70.

69. Gessen, *Istoriia zaonkodatel'stva,* 71.

70. Gessen, *Istoriia zaonkodatel'stva,* 71.

71. Gessen, *Istoriia zaonkodatel'stva,* 71.

72. Gessen, *Istoriia zaonkodatel'stva,* 71.

73. Gessen, *Istoriia zaonkodatel'stva,* 70.

74. Zelnik's edited volume has an invaluable discussion of the intelligentsia's perceptions of "an ideal" worker. See Zelnik, *Workers and Intelligentsia.*

75. Zelnik, *Workers and Intelligentsia,* 70–71.

76. Pankratova, *Istoriia proletariata,* 60.

77. "Bolezni rabochikh," *Znanie* 6 (1872): 51.

78. Here the author refers to N. Flerovskii, a pseudonym for Vil'ghel'm Vil'gel'movich Bervi (1829–1918), a Russian sociologist, economist, publicist, theorist of Populism, and public activist during the 1860s and 1890s. He was born to a British émigré family. His father was a professor of physiology at Kazan' University. Shchapov (Afanasii Prokof'evich Shchapov) was a Siberian philosopher, historian, essayist, and writer. He became actively involved in the student movement in Kazan'. See Ivanov and Volin, *Istoriia rabochego,* 125–26.

79. Andreev, *Rabota,* 41.

80. GARF, *fond* 102, *opis'* 42, *delo* 34(1), *list* 78; TsIAM, *fond* 2354, *opis'* 1, *delo* 49, "Ustav o lichnom naime rabochikh i prislugi i zamechaniia k nemy," *listy* 1–38; Andreev, *Rabota,* 7, 41–88; P. Litvinov-Falinskii, *Fabrichnoe zakonodatel'stvo,* 13–28.

81. Gessen, *Trud,* 48.

82. GARF, *fond* 102, *opis'* 42, *delo* 34 (1), *listy* 111–15; Andreev, *Rabota,* 10.

83. Andreev, *Rabota,* 74.

84. Andreev, *Rabota,* 25–26; Gessen, *Trud,* 126.

85. Cited in Gessen, *Istoriia zakonodatel'stva,* 75–76.

86. Recent scholarship on imperial Russia suggests that a civil society, as embodied in voluntary activism and numerous and active public associations, had fully emerged, despite some weaknesses and shortcomings. For a helpful theoretical discussion of civil society in imperial Russia, see Bradley, "Subjects into Citizens." For specific case studies on the subject, see Melancon, *Lena Goldfields Massacre;* Lindenmeyer, *Poverty Is Not a Vice;* Wirtschafter, *Structures of Society;* Wagner, *Marriage.* Elements of civil society and civic consciousness were emerging even during prereform decades. For a pertinent discussion of prereform civic culture, see Gorshkov, "Democratizing Habermas."

87. Andreev, *Rabota,* 42.

88. Andreev, *Rabota,* 42–43.

89. Gessen, *Istorria zakonodatel'stva*, 79. For more discussion of the Krenholm strike, see Zelnik, *Law and Disorder*.

90. Andreev, *Rabota*, 50.

91. Andreev, *Rabota*, 50.

92. Andreev, *Rabota*, 43–64.

93. Andreev, *Rabota*, 65.

94. Andreev, *Rabota*, 64–67. For a discussion of European child labor laws, see Gorshkov, "Toward a Comprehensive Law," 62.

95. GARF, *fond* 102, *opis'* 42, *delo* 34 (2), *list* 25.

96. Litvinov-Falinskii, *Fabrichnoe zakonodatel'stvo*, 13.

97. Andreev, *Rabota*, 9.

98. Gessen, *Istoriia zakonodatel'stva*, 78.

99. Andreev, *Rabota*, 33–37.

100. Gessen, *Istoriia zakonodatel'stva*, 78.

101. Andreev, *Rabota*, 63–64, 70, 72.

102. Quoted in Gessen, *Trud*, 48–49.

103. *Arkhiv sudebnoi meditsiny i obshchestvennoi gigieny* (The Archive of Criminal Medicine and Public Hygiene) 1 (1871): 127.

104. *Arkhiv*, 139–40.

105. *Znanie* 6 (1872): 49.

106. Zhuk, *Razvitie*, 321.

107. See, e.g., the 1870s issues of *Vestnik Evropy*.

108. Liadov, *Rukovodstvo*.

109. Zhuk, *Razvitie*, 303.

110. Chekhov, *Izbrannye proizvedenia*, 1: 423.

111. Chekhov, *Izbrannye proizvedenia*, 1: 250–53.

112. Setin, "Rozhdeniie," 128–29.

113. See, e.g., Chekhov's "Zhizn' v voprosakh i vosklitsaniiakh" and "Sbornik dlia detei."

114. GARF, *fond* 102, *opis'* 42, *delo* 34(2), *list* 26.

Chapter 4

1. *Sbornik postanovlenii*; Mikulin, *Fabrichnaia inspektsiia*, 10.

2. *Sbornik postanovlenii* 5, no. 3.

3. Balabanov, *Nashi zakony*, 7.

4. Balabanov, *Nashi zakony*, 9–10.

5. Mikulin, *Fabichnaia inspektsiia*, 10–11; Mikhailovskii, *O deiatel'nosti*, 9–10, 12.

6. *Sbornik postanovlenii* 4, no. 6, and 12, no. 16.

7. Siberian mining and metallurgical industries were regulated by earlier laws introduced during the 1830s and 1840. These laws provided for a system of inspectors who supervised work-related issues. For a discussion of eastern Siberia, see Melancon, *Lena Goldfields Massacre*, 34–35.

8. *Sbornik postanovlenii*, 24.

9. "Vnutrennee obozrenie," *Vestnik Evropy* 4 (Aug. 1882): 724; Mikulin, *Fabrichnaia inspektsiia*, 8; *Sbornik postanovlenii* 12, no. 3.

10. In 1886, fourteen factory inspectors and their assistants were engineers, ten were

medical doctors, and three were educators and university professors (Mikhailovskii, *O deiatel'nosti*, 5).

11. "Vnutrennee obozrenie," *Vestnik Evroy* 6 (Dec. 1886): 861.
12. "Vnutrennee obozrenie," *Vestnik Evroy* 6 (Dec. 1886): 865–66.
13. Gessen, *Istoriia zakonodatel'stva*, 86–88.
14. Gessen, *Istoriia zakonodatel'stva*, 85–86.
15. Gessen, *Istoriia zakonodatel'stva*, 86. Nikolai Khristophorovich Bunge, a liberal political economist and doctor of political science, was the finance minister from 1881 to 1886.
16. Mikulin, *Fabrichnaia inspektsiia*, 10.
17. Mikulin, *Fabrichnaia inspektsiia*, 10–11; *Sbornik postanovlenii*, app. "Pravila," no. 1.
18. Mikhailovskii, *O deiatel'nosti*, 8.
19. "Vnutrennee obzrenie," *Vestnik Evropy* 6 (Dec. 1886): 867.
20. "Vnutrennee obzrenie," *Vestnik Evropy* 6 (Dec. 1886): 867.
21. Mikhailovskii, *O deiatel'nosti*, 8, 13.
22. "Vnutrennee obozrenie," *Vestnik Evropy* 6 (Dec. 1886): 865; Mikhailovskii, *O deiatel'nosti*, 13.
23. *Sbornik postanovlenii*.
24. "Vnutrennee obozrenie," *Vestnik Evropy* 4 (Aug. 1882): 722–26; "Iz obshchestvennoi khroniki," *Vestnik Evropy* 4 (July 1884): 447–48.
25. Mikhailovskii, *O deiatel'nosti*, 7–8.
26. Mikhailovskii, *O deiatel'nosti*, 4.
27. Mikhailovskii, *O deiatel'nosti*, 7–8; *Sbornik postanovlenii*, app. "Pravila."
28. Mikulin, *Fabrichnaia inspektsiia*, 10.
29. "Vnutrennee obozrenie," *Vestnik Evropy* 4 (Aug. 1882): 722.
30. See letters of the minister of the interior, Count D. A. Tolstoi, to the ministers of finance and justice (Feb. 4, 1885) (GARF, *fond* 102, *opis'* 42, *delo* 34 [1], *listy* 1–3, 4–5). For further discussion of workers' unrest during the period, see Laverychev, *Tsarism*, 14.
31. Cited in Kirov, *Na zare*, 83. On workers' protests during the late 1870s and early 1880s, see Ivanov and Volin, *Istoriia rabochego*, 126, 127–28.
32. *Sbornik postanovlenii*, 18–22, and app. "Spisok."
33. *Sbornik postanovlenii*, 6–10; Mikhailovskii, *O deiatel'nosti*, 6–14; Balabanov, *Nashi zakony*, 39.
34. *PSZ*, 3rd ser. (hereafter *PSZ* 3), no. 3013.
35. Mikulin, *Fabrichnaia inspektsiia*, 11–12; Ivanov and Volin, *Istoriia rabochego*, 124.
36. Litvinov-Falinskii, *Fabrichnoe zakonodatel'stvo*, 314–18.
37. GARF, *fond* 102, *opis'* 42, *delo* 34 (15), *list* 116.
38. *PSZ* 3, no. 6741.
39. *PSZ* 3, nos. 8402, 11391. Also see *Svod zakonov Rossiiskoi Imperii* (hereafter *SZ*) (St. Petersburg, 1893), vol. 7, *Ustav gornyi*, no. 655.
40. Mikulin, *Fabrichnaia inspektsiia*, 14–15.
41. Mikulin, *Fabrichnaia inspektsiia*, 13–14.
42. *Rabochii vopros*, 41, no. 5.
43. *Rabochii vopros*.
44. *SZ*, vol. 21, part 2, *Ustav o promyshlennom trude* (St. Petersburg, 1913).
45. *Rabochii vopros*, nos. 15, 25, 27; *PSZ* 3 (1905), no. 26987, see sect. 2, p. 852; *SZ* (1913), vol. 21, part 2, *Ustav o promyshlennom trude*, no. 230.

46. For discussion, see Freidgut, *Iuzovka and Revolution*.

47. Pazhitnov, *Polozhenie*, 146–48; Arutiunov, *Rabochee dvizhenie*, 258.

48. *Svod Zakonov*, vol. 21, part 2, *Ustav o promyshlennom trude*; Gorshkov, "Toward a Comprehensive Law," 64.

49. Balabanov, *Nashi zakony*.

50. See, e.g., numerous issues of *Vestnik Evropy* that gave labor laws broad public discussion: 11 (1884), 2 (1885), 6 (1886).

51. Hepple, *Making of Labor Law*, chap. 2.

52. For discussion, see Hepple, *Making of Labor Law*, chap. 2.

53. "Vnutrennee obozrenie," *Vestnik Envropy* 4 (Aug. 1882): 722.

54. For discussion, see Nardinelli, *Child Labor*, 144.

55. See, e.g., Jenson, "Representation of Gender"; Lewis and Rose, "Let England Blush." Also see Kessler-Harris, Lewis, and Wikander, "Introduction."

56. For discussion, see Tuttle, *Hard at Work*, 228. Late imperial Russian scholars of child labor have also argued that child labor laws led to the decline of children's employment in industries.

57. Gorshkov, "Toward a Comprehensive Law," 50; Lenin, "Novyi fabrichnyi zakon" (1897).

58. Mikhailovskii, *O deiatel'nosti*, 96.

59. Mikhailovskii, *O deiatel'nosti*, 38.

60. "Vnutrenee obozrenie," *Vestnik Evropy* 6 (Dec. 1886): 862.

61. Mikulin, *Fabrichnaia inspektsiia*, 14–15.

62. Mikhailovskii, *O deiatel'nosti*, 15.

63. "Vnutrennee obozrenie," *Vestnik Evropy* 6 (Dec. 1886): 863.

64. Peskov, *Fabrichnyi byt*; Peskov, *Otchet fabrichnogo inspektora*, 25–26.

65. Gessen, *Istoriia zakonodatel'stva*, 100.

66. Gessen, *Istoriia zakonodatel'stva*, 100.

67. Peskov, *Fabrichnyi byt*; Peskov, *Otchet fabrichnogo inspektora*, 25–26.

68. Gessen, *Trud*, 65.

69. Cited in Pazhitnov, *Polozhenie*, 2: 28.

70. Peskov, *Fabrichnyi byt*, 12–13.

71. *Za nevskoi*, 15.

72. Peskov, *Fabrichnyi byt*, 12–13, 18.

73. GARF, *fond* 102, *opis'* 42, *delo* 34 (15), *listy* 79, 85. See also Gessen, *Trud*, 111.

74. GARF, *fond* 102, *opis'* 42, *delo* 34 (15), *listy* 1–132.

75. *Svod otchetov*. See also Rashin, *Formirovanie*.

76. Semanov, *Peterburgskie*.

77. GARF, *fond* 533, *opis'* 1, *delo* 457. For short bibliographies of worker activists, see Pirani, *Russian Revolution*, app. 1 (243–56).

78. Cited in Kirov, *Na zare*, 60.

79. Mikhailovskii, *O deiatel'nosti*, 65.

80. *Za nevskoi*, 17

81. Mikulin, *Prichiny*, prilozheniia, 7.

82. Gessen, *Zakonodatel'stvo*, 100.

83. Kirov, *Na zare*, 65; Zaitsev, *Polozhenie*, 16.

84. Mikhailovskii, *O deiatel'nosti*, 86.

85. Mikhailovskii, *O deiatel'nosti*, 87–88.

86. Balabanov, *Ocherki*, 3: 150; Gessen, *Trud*, 258. The figures on the wages for 1905 and 1910 are cited in *Rabochee dvizhenie*, 50.

87. Gessen, *Trud*, 250–51.

88. Dement'ev, *Fabrika*, 181.

89. Gessen, *Trud*, 64–111.

90. Gessen, *Trud*, 111.

91. "Biulleten' Ministerstva Truda" (June 10, 1917), GARF, *fond* 4100, *opis'* 1, *delo* 69 (1917), *list* 3; Gessen, *Trud*, 111.

92. A. Anikst, "Edinstvo organov po uchetu i raspredeleniiu rabochei sily," *Ekonomicheskaia zhizn'*, June 29, 1919.

93. For discussion of this issue, see Ball, *And Now My Soul*; Goldman, *Women, the State*, chap. 2; Stolee, "Homeless Children." For a more recent interpretation, see Gorshkov, "Children's Commission."

94. Kitanina, *Rabochie*, 176.

95. Gorshkov, *Life under Russian Serfdom*, 54–57.

96. Gorshkov, "Serfs on the Move," 652–63.

97. This idea has been emphasized by some British scholars of education. See Kirby, *Child Labor*, 114.

98. For further discussion of these early educational establishments, see Koshman, "Fabrichnye shkoly."

99. Koshman, "Fabrichnye shkoly," 23; Laverychev, *Tsarism*, 16.

100. "Voskresnye shkoly v Moskve," *Ekonom* 22 (Moscow, 1846), 166.

101. RGIA, *fond* 560, *opis'* 38 (1844), *delo* 499, *list* 23.

102. RGIA, *fond* 560, *opis'* 38 (1846), *delo* 530, *list* 112.

103. TsIAM, *fond* 17, *opis'* 97 (1845), *delo* 81, *list* 395.

104. The issue of rural schools after 1861 is thoroughly discussed in Ekloff, *Russian Peasant Schools*.

105. Andreev, *Rabota*, 172.

106. Mikhailovskii, *O deiatel'nosti*, 86.

107. Mikhailovskii, *O deiatel'nosti*, 89; Ivanov and Volin, *Istoriia rabochego*, 126.

108. Peskov, *Fabrichnyi byt*, 26.

109. RGIA, *fond* 560, *opis'* 38 (1846), *delo* 530, *list* 112.

110. Mikhailovskii, *O deiatel'nosti*, 92.

111. Ivanov and Volin, *Istoriia rabochego*, 127.

112. Ivanova, *Promyshlennyi*, 266.

113. Ivanova, *Promyshlennyi*, 266.

114. Ivanova, *Promyshlennyi*, 266.

115. Melancon, *Lena Goldfields Massacre*, 59–60.

116. "Obshchestvo 'Detskii Trud i otdykh," in *Biblioteka novogo vospitania i obrazovaniia i zashchity detei*, vyp. 28, 1–8 (cited on 2).

117. Arutiunov, *Rabochee dvizhenie*, 41.

118. Melancon, *Lena Goldfields Massacre*, 87.

119. Cited in Melancon, *Lena Goldfields Massacre*, 59.

120. Kirov, *Na zare*, 94.

121. Kabanov, Erman et al., *Ocherki*, 54.

122. Gessen, *Istoriia zakonodatel'stva*, 79.

123. Cited in Kirov, *Na zare*, 100; *Iskra* 14 (1902).

124. *Rabochee dvizhenie*, nos. 68, 69, 70; RGIA, *fond* 1405 (Ministerstvo iustitsii), *opis'* 104, *delo* 3315, *listy* 4–4a.

125. *Rabochee dvizhenie*, no. 29.

126. *Iskra* 85 (1905).

127. GARF, *fond* 533, *opis'* 1, *delo* 96, *list* 17.

128. Kirov, *Na zare*, 96–99.

129. Kirov, *Na zare*, 99.

130. Kirov, *Na zare*, 102–3.

131. Kirov, *Na zare*, 84.

132. Kirov, *Na zare*, 106.

133. For an informative discussion of this issue, see Herrlinger, *Working Souls*.

134. The Bund was the All-Jewish Union of Workers. Organized in 1897, it became part of the Communist Party in 1921 (Kirov, *Na zare*, 124).

135. Cited in Kirov, *Na zare*, 110.

136. GARF, *fond* 533, *opis'* 5, *delo* 95, *list* 81 ob.

137. Kirov, *Na zare*, 112.

138. GARF, *fond* 533, *opis'* 1, *delo* 29, *list* 2.

139. Pirani, *Russian Revolution*, 246.

140. Roseman, *Generations in Conflict*; Esler, *Youth Revolution*; Gillis, *Youth and History*; Heer, *Challenge of Youth*; Feuer, *Conflict of Generations*.

141. Cited in Kirov, *Na zare*, 124.

142. GARF, *fond* 533, *opis'* 1, *delo* 452, *list* 4.

143. For this interpretation of late imperial Russian political culture, see Melancon, "Popular Political Culture"; Melancon and Pate, "Bakhtin Contra Marx and Lenin."

Conclusion

1. Gorshkov, "Toward a Comprehensive Law", 56, 62–64.

2. Wortman, *Scenarios of Power*, contains the classic statement of Russian autocracy's self-image.

3. Wirtschafter, *Social Identity*, stresses the absence of intermediary societal entities, whereas other studies, including McCaffrey and Melancon, *Russia in the European Context*, and Bradley, *Voluntary Associations*, draw attention to intermediate cultural and social phenomena.

4. For a discussion, see Neuberger, *Hooliganism*.

5. For the historiography of children's homelessness and neglect, see Gorshkov, "Children's Commission."

Appendix

1. *PSZ* 2, vol. 20, no. 19262, 591.

2. *Sbornik postanovlenii*.

3. TsIAM, *fond* 32, *opis'* 1857, *delo* 2274, *list* 6.

4. TsIAM, *fond* 32, *opis'* 1857, *delo* 2274, *list* 11.

5. TsIAM, *fond* 32, *opis'* 1857, *delo* 2274, *list* 12.

Bibliography

Primary Archival Sources

Gosudarstvennyi arkhiv Rossiiskoi Federatsii (GARF) [State Archive of the Russian Federation]. Moscow.

   fond 102: Departament politsii [Police Department]
   fond 109: Tret'e otdelenie sekretnoi Ego Imperatorskogo Velichestva Kantseliarii [The Third Department]
   fond 533: Vsesoiuznoe obshchestvo politkatarzhan i ssyl'noposelentsev [Soviet Society of Political Prisoners]
   fond 4100: Ministerstvo truda Vremennogo Pravitel'stva [Labor Ministry of the Provisional Government]

Rossiiskii gosudarstvennyi arkhiv drevnikh aktov (RGADA) [Russian State Archive of Ancient Acts]. Moscow.

Rossiiskii gosudarstvennyi istoricheskii arkhiv (RGIA) [Russian State Historical Archive]. St. Petersburg.

   fond 1405: Ministerstvo iustitsii [Ministry of Justice]

Tsentral'nyi istricheskii arkhiv goroda Moskvy (TsIAM) [Central Historical Archive of Moscow]. Moscow.

   fond 16: Kantseliariia moskovskogo general-gubernatora [Office of Moscow Military Governor]
   fond 17: Kantseliariia moskovskogo grazhdanskogo gubernatora [Office of Moscow Civic Governor]
   fond 1780: Komissiia dlia osmotra fabrik i zavodov v Moskve [Commission for Inspection of Factories and Workshops in Moscow]
   fond 2354: Mockovskoe otdeleniie Manufakturogo Coveta [Moscow Section of the Manufacturing Council]

Periodicals and Other Primary Sources

Andreev, E. N. *Rabota maloletnikh v Possii i zapadnoi Evrope*, vol. 1. St. Petersburg, 1884.
Arkhiv sudebnoi meditsiny i obshchestvennoi gigieny. St. Petersburg.

Biblioteka dlia chteniia. St. Petersburg.

Gmelin, Johann Georg. *Reise durch Sibirien von dem Jar 1733 bis 1743.* Gottingen: Velegts Anram, 1751–52.

Ianzhul, I. I. *Fabrichnyi byt Moskovskoi gubernii. Otchet za 1882–1883 gg. fabrichnogo inspectora.* St. Petersburg: Tip. Kirshbauma, 1884.

Liadov, V. I. *Rukovodstvo k vospitaniiu detei.* St. Petersburg: Tip. Kolesova i Mikhina, 1873.

Mikhailovskii, I. T. *O deiatel'nosti fabrichnoi inspektsii: Otchet za 1885 god glavnogo fabrichngo inspektora I. T. Mikhailovskogo.* St. Petersburg: Tip. B. Kirshbauma, 1886.

Pallas, P. S. *Puteshestviia po raznym provintsiiam Rossiiskago gosudarstva.* St. Petersburg, 1788.

Peskov, P. A. *Fabrchnyi byt Vladimirskoi gubernii: Otchet za 1882–1883 god fabrichnago inspektora za zaniatiiami malometnikh rabochikh Vladimirskago okruga P. A. Peskova.* St. Petersburg: Tip. B. Kirshbauma, 1884.

———. *Otchet fabrichnogo inspektora Vladimirskogo okruga P. A. Peskova za 1885 g.* St. Petersburg, 1886.

Polnoe sobranie zakonov Rossiiskoi Imperii. 1st series, 1649–1824. St. Petersburg: Tip. 2-go otdeleniia EIV Kentseliarii, 1830.

Polnoe sobranie zakonov Rossiiskoi Imperii. 2nd series, 1826–1879. St. Petersburg: Tip. 2-go otdeleniia EIV Kentseliarii, 1885.

Proekt pravil dlia fabrik i zavodov v S.-Peterburge i uezde. St. Petersburg, 1860.

Protokoly i stengraficheskie otchety 1-go Vserossiiskogo S'ezda fabrikantov, zavodchikov i lits interesuiushchikhcia otechestvennoi promyshlemmost'iu, v 1870 g. St. Petersburg: Izd. Akademii Nauk, 1872.

Rabochee dvizhenie v Rossii v XIX veke. Sbornik dokumentov i materialov. Ed. A. M. Pankratova. 4 vols. Moscow: Gos. izd-vo polit. lit-ry, 1950–63.

Rabochee dvizhenie v Rossii v XIX veke. Sbornik dokumentov i materialov. Ed. A. M. Pankratova. 2nd ed. Moscow: Gos. izd-vo polit. lit-ry, 1955–.

Rabochee dvizhenie v Rossii v 1901-1904 gg. Sbornik dokumentov. Ed. L. M. Ivanov, I. M. Pushkareva et al. Leningrad: Nauka, Leningr. otd., 1975.

Rabochee dvizhenie v Rossii v 1901-1904 gg. Sbornik dokumentov. Ed. V. I. Vel'bel', N. N. Dorzhiev, and N. A. Mal'tseva. Leningrad: Nauka, Leningr. otd-nie, 1975.

Rabochee dvizhenie v Rossii v period revoliutsii 1905–1907 gg. Materialy dlia "Khroniki rabochego dvizheniia." Ed. IU. I. Kir'ianov et al. Moscow: Glavnoe arkhivnoe upravlenie pri Kabinete Ministrov SSSR, 1991.

Rabochii vopros v komissii V. N. Kokovtsova v 1905 gody. Moscow, 1926.

Sbornik postanovlenii o maloletnikh rabochikh na zavodakh, fabrikakh i v drugikh promyshlennykh zavedeniiakh. St. Petersburg, 1885.

Svod otchetov fabrichnykh inspektorov za 1914 g. Petrograd, 1915.

Trudy komissii uchrezhdennoi dlia peresmotra ustavov fabrichnogo i remeslennogo. St. Petersburg, 1863.

Vestnik Evropy. Moscow.

Za nevskoi Zastavoi. Zapiski rabochego Alekseia Buzinova. Moscow and Leningrad: Gosizdatel'stvo, 1930.

Zhurnal Russkogo obshchestva okhraneniia narodnogo zdraviia.

Secondary Literature

Aleksandrov, V. A. *Sel'skaia obshchina v Rossii, XVII–nachalo XIX v.* Moscow: Nauka, 1976.

Anokhina, L. A., and M. N. Shmeleva. *Byt gorodskogo naseleniia stednei polosy RSFSR v proshlom i nastoiashchem.* Moscow: Nauka, 1977.

Aries, Philippe. *Centuries of Childhood: A Social History of Family Life.* Trans. R. Baldick. New York: Knopf, 1962.

Arutiunov, G. A. *Rabochee dvizhenie v Rossii v period novogo revoliutsionnogo pod"ema, 1910–1914.* Moscow: Nauka, 1975.

Baklanova, E. N. [Elena Nikolaevna]. *Krest'ianskii dvor i obshchina na Russkom Severe, konets XVII–nachalo XVIII v.* Moscow: Nauka, 1976.

Balabanov, M. *Nashi zakony o zashchite fabrichnogo truda detei. Obshchedostupnoe prakticheskoe rukovodstvo.* Kiev: Tip. pervoi kievskoi arteli pechatnikov, 1915.

———. *Ocherki po istorii rabochego klassa v Rossii.* Kiev: Sorabkop, 1924.

Ball, Alan. *And Now My Soul Is Hardened: Abandoned Children in Soviet Russia.* Berkeley: University of California Press, 1994.

Bernshtam, T. A. *Molodezh v obriadovoi zhizni russkoi obshchiny XIX–nachala XX v.* Leningrad: Nauka, 1988.

Blackwell, William L. *The Beginnings of Russian Industrialization, 1800–1860.* Princeton: Princeton University Press, 1968.

Blum, Jerome. *Lord and Peasant in Russia from the Ninth to the Nineteenth Century.* Princeton: Princeton University Press, 1961.

Bradley, Joseph. "Subjects into Citizens: Societies, Civil Society, and Autocracy in Tsarist Russia." *American Historical Review* 107, no. 4 (2002): 1094–23.

———. *Voluntary Associations in Tsarist Russia: Science, Patriotism, and Civil Society.* Cambridge: Harvard University Press, 2009.

Brandenburgskii, P. "Zheleznye zavody v Tul'skom, Kashirskom and Aleksinskom uezdakh v XVII stoletii." In *Oruzheinyi Sbornik*, books 1–4. St. Petersburg: Tip. Artill. Zhurnala, 1875.

Brower, Daniel. *The Russian City between Tradition and Modernity, 1850–1900.* Berkeley: University of California Press, 1990.

Bukhina, V. "Iz istorii detskogo truda v krepostnoi Rossii." In *Istoriia proletariata SSSR*, ed. A. M. Pankratova, 4: 117–49. Moscow: Kommunisticheskaia Akademiia, 1930.

———. "Maloiaroslavetskaia fabrika do zakreposhcheniia, 1718–1737." In *Istoriia proletariata SSSR*, ed. A. M. Pankratova, 2: 115–47. Moscow: Kommunisticheskaia Akademiia, 1930.

Burbank, Jane. *Russian Peasants Go to Court: Legal Culture in the Countryside, 1905–1917.* Bloomington: Indiana University Press, 2004.

Burbank, Jane, Mark Von Hagen, and Anatolii Remnyev, eds. *Russian Empire: Space, People, Power, 1700–1930.* Bloomington: Indiana University Press, 2007.

Burds, Jeffrey. *Peasant Dreams and Market Politics: Labor Migration and the Russian Village, 1861–1905.* Pittsburgh: University of Pittsburgh Press, 1998.

Chebotarev, Kh. *Istoricheskoe i topographicheskoe opisanie gorodov Moskovskoi gubernii s ikh uezdami.* Moscow, 1787.

Chekhov, A. P. *Izbrannye proizvedeniia v trekh tomakh.* 3 vols. Moscow: Izd. Khudozhestvennaia literatura, 1967.

Crisp, Olga. *Studies in the Russian Economy before 1914*. New York: Barnes and Noble Books, 1976.

Dement'ev, E. M. *Fabrika, chto ona daet naseleniiu i chto ona u nego beret*. Moscow: Izd. D. I. Sytina, 1897.

Economakis, Evel G. "Patterns of Migration and Settlement in Revolutionary St. Petersburg: Peasants from Iaroslavl' and Tver Provinces." *Russian Review* 56 (January 1997): 8–24.

Ekloff, Ben. *Russian Peasant Schools: Officialdom, Village Culture, and Popular Pedagogy*. Berkeley: University of California Press, 1986.

Enalieva, N. E. "Child Labor and Industrial Apprenticeship in Russia." *Russian Education and Society* (November 1995): 11–18.

Engel, Barbara Alpern. *Between the Fields and the City: Women, Work and the Family in Russia, 1861–1914*. Cambridge: Cambridge University Press, 1993.

Engelstein, Laura. "Combined Underdevelopment: Discipline and the Law in Imperial and Soviet Russia." *American Historical Review* (April 1993): 338–81.

Esler, Anthony, ed. *The Youth Revolution: The Conflict of Generations in Modern History*. Lexington, Mass.: D. C. Heath, 1974.

Feuer, Lewis S. *The Conflict of Generations*. New York: Basic Books, 1969.

Freidgut, Theodore H. *Iuzovka and Revolution*. Vol. 1, *Life and Work in Russia's Donbass, 1869–1924*. Vol 2, *Politics and Revolution in Russia's Donbass, 1869–1924*. Princeton: Princeton University Press, 1989, 1994.

Frierson, Cathy A. "Razdel: The Peasant Family Divided." In *Russian Peasant Women*, ed. Beatrice Farnsworth and Lynne Viola, 73–88. New York: Oxford University Press, 1992.

Fuller, Raymond. "Child Labor." In *Encyclopedia of the Social Sciences*, ed. E. R. A. Seligman. New York: Macmillan, 1930.

Gerasimov, V. G. *Zhizn' russkogo rabochego: Vospominaniia*. Moscow, 1959.

Gessen, V. Iu. *Istoriia zakonodatel'stva o trude rabochei molodezhi v Rossii*. Leningrad: Izdatel'stvo leningradskogo Gubprofsoveta, 1927.

———. *Trud detei i podrostkov v Rossii. Ot XVII veka go Oktiabr'skoi Revoliutsii*. Moscow and Leningrad: Gosudarstvennoe izdatel'stvo, 1927.

Giffen, Frederick C. "The 'First Russian Labor Code': The Law of June 3, 1886." *Russian History* 2 (1975): 83–102.

Gillis, John R. *Youth and History. Tradition and Change in European Age Relations, 1770–Present*. New York: Academic Press, 1974.

Glickman, Rose B. *Russian Factory Women: Workplace and Society, 1880–1914*. Berkeley: University of California Press, 1984.

Goldman, Wendy Z. *Women, the State and Revolution. Soviet Family Policy and Social Life, 1917–1936*. Cambridge and New York: Cambridge University Press, 1993.

Gorovoi, F. S. *Otmena krepostnogo prava i rabochie volneniia na Urale i v Permskoi gubernii*. Molotov: Molotovskoe knizhnoe izd-vo, 1954.

———. *Padenie krepostnogo prava na gornykh zavodakh Urala*. Perm': Permskoe knizhnoe izd-vo, 1961.

Gorshkov, Boris B. "Children's Commission." Supplement to the *Modern Encyclopedia of Russian, Soviet and Eurasian History*, 15–18, 2005.

———. "Democratizing Habermas: Peasant Public Sphere in Pre-Reform Russia" *Russian History/Histoire Russe* 31 (Winter 2004): 373–85.

———. "Factory Children: An Overview of Child Industrial Labor and Laws in Imperial Russia, 1840–1914." In *New Labor History: Worker Identity and Experience in Russia, 1840–1918*, ed. Michael Melancon and Alice K. Pate, 9–33. Bloomington: Slavica, 2002.

———, ed. and trans. *A Life under Russian Serfdom: The Memoirs of Savva Dmitrievich Purlevskii, 1800–1868*. Budapest and New York: Central European University Press, 2005.

———. "Serfs on the Move: Peasant Seasonal Migration in Pre-Reform Russia, 1800–61." *Kritika*, new series, 1 (Fall 2000): 627–56.

———. "Toward a Comprehensive Law: Tsarist Factory Labor Legislation in European Context, 1830–1914." In *Russia in the European Contest, 1789–1914: A Member of the Family*, ed. Susan P McCaffray and Michael Melancon, 49–70. New York: Palgrave MacMillan, 2005.

Gromyko, M. M. *Mir russkoi derevni*. Moscow: Molodaia gvardiia, 1991.

———. *Traditsionnye normy povedeniia i formy obshcheniia russkikh krest'ian XIX veka*. Moscow: Nauka, 1986.

———. *Trudovye traditsii krest'ian Sibiri, XVIII—pervaia polovina XIX v*. Novosibirsk: Nauka, 1975.

Habermas, Juergen. *The Philosophical Discourse of Modernity: Twelve Lectures*. Trans. Frederick Lawrence. Cambridge: MIT Press, 1987.

———. *The Structural Transformation of the Public Sphere: An Inquiry into a Category of Bourgeois Society*. Trans. Thomas Burger and Frederick Lawrence. Cambridge: MIT Press, 1989.

Hammond, J. L., and B. Hammond. *The Town Labourer, 1760–1832: The New Civilization*. London: Longman's, 1966.

Hartwell, R. M. *The Industrial Revolution and Economic Growth*. London: Methuen, 1971.

Heer, Friedrich. *Challenge of Youth*. Tuscaloosa: University of Alabama Press, 1974.

Hepple, Bob, ed. *The Making of Labor Law in Europe: A Comparative Study of Nine Countries up to 1945*. London and New York: Mansell Publishing, Ltd., 1986.

Herrlinger, Page. *Working Souls: Russian Orthodoxy and Factory Labor in St. Petersburg, 1881–1917*. Bloomington: Slavica, 2007.

Hetherington, E. Mavis, et al., eds. *Child Development in Life-Span Perspective*. Hillsdale, N.J.: Lawrence Erlbaum Associates, 1988.

Heywood, Colin. *Childhood in Nineteenth-Century France: Work, Health, and Education among the Classes Populaires*. Cambridge and New York: Cambridge University Press, 1988.

———. *History of Childhood: Children and Childhood in the West from Medieval to Modern Times*. Malden, Mass.: Blackwell, 2001.

Hindman, Hugh D. *Child Labor: An American History*. New York: M. E. Sharpe, 2002.

Hobbs, Sandy, Jim McKechnie, and Michael Lavalette, eds. *Child Labor: A World History Companion*. Santa Barbara: ABC-CLIO, 1999.

Horn, Pamela. *The Victorian and Edwardian Schoolchild*. Gloucester: Sutton, 1989.

Horn, Pamela, et al., eds. *Children's Work and Welfare, 1780–1890*. Cambridge and New York: Cambridge University Press, 1995.

Horrell, Sara, and Jane Humphries, "'The Exploitation of Little Children': Child Labor and the Family Economy in the Industrial Revolution." *Explorations of Economic History* 32 (1995): 485–516.

Isaev, G. S. *Rol' tekstil'noi promyshlennosti v genezise i razvitii kapitalizma v Rossii, 1760–1860*. Leningrad: Nauka, 1970.

Ivanov, L. M., and M. S. Volin, eds. *Istoriia rabochego classa Rossii, 1861–1900*. Moscow: Nauka, 1972.

Ivanova, N. A. *Promyshlennyi tsentr Rossii, 1907–1914*. Moscow: Inst. rossiiskoi istorii RAN, 1995.

Jenson, Jane. "Representation of Gender: Policies to 'Protect' Women Workers and Infants in France and the United States before 1914." In *Women, the State and Welfare*, ed. Linda Gordon, 152–98. Madison: University of Wisconsin Press, 1991.

Johnson, Robert E. *Peasant and Proletarian: The Working Class of Moscow in the Late Nineteenth Century*. New Brunswick: Rutgers University Press, 1979.

Kabanov, P. I., P. K. Erman et al., eds. *Ocherki istorii rossiiskogo proletariata*. Moscow: Izdatel'stvo sotsial'no-ekonomicheskoi literatury, 1963.

Kabuzan, V. M. *Russkie v mire: dinamika chislennosti i rasseleniia (1719–1989). Formirovaniie eticheskikh i politicheskikh granits russkogo naroda*. St. Petersburg: Russko-baltiiskii informatsionnyi tsentr, 1996.

Kahan, Arcadius. *Russian Economic History: The Nineteenth Century*. Ed. Roger Weiss. Chicago: University of Chicago Press, 1985.

Kazantsev, B. N, *Rabochie Moskvy i Moskovskoi gubernii v seredine XIX veka*. Moscow: Nauka, 1976.

Kelly, Catriona. *Children's World: Growing Up in Russia, 1890–1991*. New Haven and London: Yale University Press, 2007.

Kessler-Harris, Alice, Jane Lewis, and Ulla Wikander. "Introduction." In Kessler-Harris, Lewis, and Wikander, *Protecting Women*, 1–28.

———, eds. *Protecting Women: Labor Legislation in Europe, the United States and Australia, 1800–1920*. Urbana and Chicago: University of Illinois Press, 1995.

Khromov, P. A. [Pavel Alekseevich]. *Ocherki ekonomiki dokapitalisticheskoi Rossii*. Moscow: Nauka, 1988.

———. *Ocherki ekonomiki tekstil'noi promyshlennosti SSSR*. Moscow and Leningrad: Izd-vo Akademii nauk SSSR, 1946.

Kirby, Peter. *Child Labor in Britain, 1770–1870*. New York: Palgrave MacMillan, 2003.

Kir'ianov, Iu. I. *Zhiznennyi uroven' rabochikh Rossii (kinets XIX–nachalo XX v.)*. Moscow: Nauka, 1979.

Kir'ianov, Iu. I., and S. I. Potolov, eds. *Rabochee dvizhenie v Rossii 1895–1904 gg.: sbornik statei i materialov dlia "Khroniki rabochego dvizheniia v Rossii v 1895–fevrale 1917 g."* Moscow: Institut istorii SSSR AN SSSR, 1988.

Kirov, A. *Na zare iunosheskogo dvizheniia v Rossii*. Chast' l. Khar'kov: Izd-vo Proletarii, Iunosheskii sektor, 1926.

Kitanina, T. M. *Rabochie peterburga v 1800–1861 gg.: Promyshlennost', Formirovanie, sostav, polozhenie rabochikh, rabochee dvizhenie*. Leningrad: Nauka, 1991.

Kleinbort, L. M. *Istoriia bezrabotitsy v Rossii: 1857–1919*. Moscow, 1925.

Koshman, Liudmila V. "Fabrichnye shkoly v Rossii v pervoi polovine XIX v." *Vestnik Moskovskogo universiteta. Istoria* 2 (1976): 20–23.

Kowler, Eileen, and Albert Martins. "Eye Movements of Preschool Children." *Science*, new series, 215 (February 19, 1982): 997–99.

Laverychev, V. A. "Iz istorii politili tsarizma po rabochemy voprosy v 60–70-e gody XIX v." *Vestnik Moskovskogo universiteta*, 3rd series (1971).

———. *Tsarism i rabochii vopros v Rossii (1861–1917 gg.).* Moscow: Mysl', 1972.
Lenin, Vladimir Il'ich. "Novyi Fabrichnyi zakon." In *Polnoe sobranie sochinenii,* by V. I. Lenin, 55 vols., 2: 39. Moscow: Izd. Politicheskoi literatury, 1967–70.
———. *Razvitie kapitalizma v Rossii.* Vol. 3 of *Polnoe sobranie sochinenii,* by V. I. Lenin, 55 vols. Moscow: Izd. Politicheskoi literatury, 1967–70.
Lewis, Jane, and Sonya Rose. "Let England Blush." In Kessler-Harris, Lewis, and Wikander, *Protecting Women,* 91–124.
Liashchenko, P. I. *Istoriia narodnogo khoziaistva SSSR.* 2 vols. 2nd ed. Moscow: Gos. izd-vo polit. lit-ry, 1956.
Lindenmeyer, Adele. *Poverty Is Not a Vice: Charity, Society, and the State in Imperial Russia.* Princeton: Princeton University Press, 1996.
Litvinov-Falinskii, P. *Fabrichnoe zakonodatel'stvo i fabrichnaia inspektsiia v Rossii.* St. Petersburg, 1900.
Martynova, A. N. "Otrazhenie deistvitel'nosti v krest'ianskoi kolybel'noi pesne. Russkii folklor. Sotsial'nyi protest v narodnoi poezii." *Russkii fol'klor* 15 (1975): 145–55
Marx, Karl. *Kapital.* Trans. Samuel Moore et al. New York: International Publishers, 1967.
Maslova, G. S. *Narodnaia odezhda v vostochnoslavianskikh traditsionnykh obychaiakh i obriadakh XIX—nachala XX veka.* Moscow: Nauka, 1984.
McCaffrey, Susan P. *The Politics of Industrialization in Tsarist Russia: The Association of Southern Coal and Steel Producers, 1874–1914.* DeKalb: Northern Illinois University Press, 1996.
———. "What Should Russia Be? Political Economy in the Thought of N. S. Mordvinov." *Slavic Review* (Autumn 2000): 572–96.
McCaffrey, Susan P., and Michael Melancon, eds. *Russia in European Context: A Member of the Family.* New York: Palgrave McMillan, 2005.
Melancon, Michael. *The Lena Goldfields Massacre and the Crisis of the Late Tsarist State.* College Station: Texas A&M University Press, 2006.
———. "Popular Political Culture in Late Imperial Russia (1800–1917)." *Russian History/Histoire Russe* 31 (Winter 2004): 369–71.
Melancon, Michael, and Alice Pate. "Bakhtin Contra Marx and Lenin: A Polyphonic Approach to Russia's Labor and Revolutionary Movements." *Russian History/Histoire Russe* 31 (Winter 2004): 387–417.
Merriman, John. *From the French Revolution to the Present.* Vol. 2 of *A History of Modern Europe.* New York: W. W. Norton, 1996.
Mikulin, A. A. [Aleksandr Aleksandrovich]. *Fabrichnaia inspektsiia v Rossiii, 1882–1906.* Kiev: Tip. Kul'zhenko, 1906.
———. *Ocherki iz istorii primeneniia zakona 3-go iiunia 1886 g. o naime rabochikh na fabrikakh i zavodakh Vladimirskoi gubernii inzhenera-mekhanika A. A. Mikulina.* Vladimir: Tip.-lit. V. A. Parkova, 1893.
———. *Prichiny i sledstviia neschastnykh sluchaev s rabochimi na fabrikakh i zavodakh. Statisticheskoe issledovanie 2543 neschastnykh sluchaev inzhenera-mekhanika A. A. Mikulina.* Odessa: Iuzhnorusskoe o-vo pechatnogo dela, 1898.
Milogolova, I. I. "Semeinye razdely v russkoi poreformennoi derevne na materialakh tsentral'nykh gubernii." *Vestnik Moskovskogo universiteta,* series 8, "History," no. 6 (1987): 37–46.
Milov, L. V. "Otkhodnichestvo." In *Sovetskaia istoricheskaia entsiklopediia,* 10: 696. Moscow: Sovetskaia entsiklopediia, 1967.

Minenko, N. A. [Nina Adamovna]. *Russkaia krest'ianskaia sem'ia v Zapadnoi Sibiri, XVIII—pervoi poloviny XIX v.* Novosibirsk: Nauka, 1979.

Minge-Kalman, Wanda. "The Industrial Revolution and the European Family: The Industrialization of Childhood as a Market for Family Labor." *Comparative Studies in Society and History* 20 (July 1978): 454–68.

Mironov, B. N. *Sotsial'naia istoriia Rossii.* 2 vols. St. Petersburg: D. Bulanin, 2000.

Moon, David. "Peasant Migration, the Abolition of Serfdom, and the Internal Passport System in the Russian Empire, c. 1800–1914." In *Coerced and Free Migration: Global Perspectives,* ed. David Eltis, 324–57. Stanford: Stanford University Press, 2002.

Nardinelli, Clark. *Child Labor and the Industrial Revolution.* Bloomington and Indianapolis: Indiana University Press, 1990.

Nekrasov, Nikolai Alekseevich. *Sobranie sochinenii v vos'mi tomakh.* 8 vols. Moscow: Khudozhestvennaia literatura, 1965.

Neuberger, Joan. *Hooliganism: Crime, Culture, and Power in St. Petersburg, 1900–1914.* Berkeley: University of California Press, 1993.

Nikol'skii, D. P. "K voprosy o vliianii fabrichnogo truda na fizicheskoe razvitie, boleznennost' i smertnost' rabochego." *Zhurnal Russkogo obshchestva okhraneniia narodnogo zdraviia* 8 (August 1895): 611–37.

Owen, Thomas C. *Capitalism and Politics in Russia: A Social History of the Moscow Merchants, 1855–1905.* Cambridge and New York: Cambridge University Press, 1981.

Pankratova, A. M. *Formirovanie proletariata v Rossii, XVII–XVIII vv.* Moskva: Izd-vo Akademii nauk SSSR, 1963.

———. *Razvitie kapitalizma v Rossii i vozniknovenie rabochego dvizhenia.* Moscow, 1947.

Paterson, Sarah V., Sabine Heim et al. "Development of Structure and Function in the Infant Brain: Implications for Cognition, and Social Behavior." *Neuroscience and Biobehavioral Reviews* 30, no. 3 (2006): 1087–1105.

Pazhitnov, K. A. *Polozhenie rabochego klassa v Possii.* 2 vols. Petrograd: Byloe, 1923.

Pirani, Simon. *The Russian Revolution in Retreat, 1920–24: Soviet Workers and the New Communist Elite.* New York: Routledge, 2008.

Post, David. *Children's Work, Schooling and Welfare in Latin America.* Boulder: Westview Press, 2001.

Rabochii klass v Rossii ot zarozhdeniia do nachala XX v. Moscow, 1983.

Rashin, A. G. [Adol'f Grigor'evich]. *Formirovanie promyshlennogo proletariata v Rossii: statistiko-ekonomicheskie ocherki.* Moscow: Gos. sotsial'no-ekonomicheskoe izd-vo, 1940.

———. *Formirovanie rabochego klassa v Rossii: Istoriko-ekonomicheskie ocherki.* Moscow, Izd-vo sotsial'no-ekon. lit-ry, 1958.

———. *Naselenie Rossii za 100 let, 1811–1913 gg.: statisticheskie ocherki.* Moscow: Gos. statisticheskoe izd-vo, 1956.

Reichman, Henry. *Railway Men and Revolution: Russia, 1905.* Berkeley: University of California Press, 1987.

Rieber, Alfred J. *Merchants and Entrepreneurs in Imperial Russia.* Chapel Hill: University of North Carolina Press, 1982.

Rodgers, Gerry, and Guy Standing, eds. *Child Work, Poverty and Underdevelopment.* Geneva: International Labor Organization, 1981.

Romanov, B. A. *Liudi i nravy Drevnei Rusi.* Moscow, 1966.

Roseman, Mark, ed. *Generations in Conflict: Youth Revolt and Generation Formation in Germany, 1770–1968*. Cambridge: Cambridge University Press, 1995.
Rudolph, Richard L. "Agricultural Structure and Proto-industrialization in Russia: Economic Development with Unfree Labor." *Journal of Economic History* 45 (March 1985): 47–69.
Ryndziunskii, P. G. [Pavel Grigor'evich]. *Kresti'iane i gorod v kapitalisticheskoi Rossii vtoroi poloviny XIX veka: vzaimootnoshenie goroda i derevni v sotsial'no-ekonomicheskom stroe Rossii*. Moscow: Nauka, 1983.
———. *Krest'ianskaia promyshlennost' v poreformennoi Rossii (60–80-e gody XIX v.)* Moscow, 1966.
———. *Utverzhdenie kapitalizma v Rossii 1850–1880 gg.* Moscow: Nauka, 1978.
Salle, Shelly. *The Whiteness of Child Labor Reform in the New South*. Athens: University of Georgia Press, 2004.
Semanov, S. N. [Sergei Nikolaevich]. *Peterburgskie pabochie nakanune pervoi russkoi revoliutsii*. Moscow and Leningrad: Nauka, 1966.
Setin, F. "Rozhdeniie prozy dlia detey." In *O literature dlia detey, Ezhegodnik*, vyp. 21. Leningrad: Izd. detskaia literatura, 1977.
Shangina, I. I. *Russkie deti i ikh igry*. St. Petersburg: Iskusstvo, 2000.
Shilov, D. N. *Gosudarstvennye deiateli rossiiskoi imperii. Glavy vysshikh i tsentral'nykh uchrezhdenin, 1802–1917. Bibliograficheskii spravochnik*. St. Petersburg: Dmitrii Bulanin, 2001.
Sovetskaia istoricheskaia entsiklopediia. 16 vols. Moscow: Sovetskaia entsiklopediia, 1967.
Stolee, Margaret K. "Homeless Children in the USSR, 1917–1957." *Soviet Studies* 40 (1988).
Sugarman, Leonie. *Life-Span Development: Concept, Theories and Interventions*. New York: Methuen, 1986.
Surh, Gerald D. *1905 in St. Petersburg: Labor, Society, and Revolution*. Stanford: Stanford University Press, 1989.
Thompson, E. P. *The Making of the English Working Class*. New York: Pantheon Books, 1963.
Tiukavkin, V. G. *Velikorusskoe krest'ianstvo i Stolypinskaia agrarnaia reforma*. Moscow: Pamiatniki istoricheskoi mysli, 2001.
Trofimov, A. S. *Rabochee dvizhenie v Rossii 1861–1894 gg.* Moscow, 1957.
Tugan-Baranovsky, Mikhail I. *Russian Factory in the Nineteenth Century*. Trans. Arthur Levin and Claora S. Levin. Homewood, Ill.: Dorsey, 1970.
———. *Russkaia fabrika v proshkom i nastoiashchem: Istoricheskoe razvitie russkoi fabriki v XIX veke*. St. Petersburg: O. N. Popovoi, 1898.
Tuttle, Caroline. *Hard at Work in Factories and Mines: The Economics of Child Labor during the British Industrial Revolution*. Boulder: Westview Press, 1999.
Vdovina, L. N. *Krest'ianskaia obshchina i monastyr' v Tsentral'noi Rossii v pervoi polovine XVIII v.* Moscow: Izdatel'stvo Moskovskogo universiteta, 1988.
Wagner, William G. *Marriage, Property and Law in Late Imperial Russia*. New York: Clarendon Press of Oxford University Press, 1994.
Walvin, James. *A Child's World: A Social History of English Childhood, 1800–1914*. Harmondsworth: Penguin Books, 1982.
Weissbach, Lee Shai. *Child Labor Reform in Nineteenth-Century France: Assuring the Future Harvest*. Baton Rouge: Louisiana State University Press, 1989.

Wirtschafter, Elise Kimerling. *Social Identity in Imperial Russia*. DeKalb: Northern Illinois University Press, 1997.

———. *Structures of Society: Imperial Russia's "People of Various Ranks."* DeKalb: Northern Illinois University Press, 1994.

Worobec, Cristine D. *Peasant Russia: Family and Community in the Post-Emancipation Period*. Princeton: Princeton University Press, 1991.

Wortman, Richard S. *The Development of a Russian Legal Consciousness*. Chicago: University of Chicago Press, 1976.

———. *Scenarios of Power: Myth and Ceremony in Russian Monarchy*. Princeton: Princeton University Press, 1995.

Zaitsev, V. A. *Polozheniie truda podrostkov i ego oplata v promyshlennosti*. Moscow: Molodaia gvardiia, 1924.

Zelnik, Reginald E. *Labor and Society in Tsarist Russia: The Factory Workers of St. Petersburg, 1855–1870*. Stanford: Stanford University Press, 1971.

———. *Law and Disorder on the Narova River: The Kreenholm Strike of 1872*. Berkeley: University of California Press, 1995.

———, ed. and trans. *A Radical Worker in Tsarist Russia: The Autobiography of Semen Ivanovich Kanatchikov*. Stanford: Stanford University Press, 1986.

———, ed. *Workers and Intelligentsia in Late Imperial Russia: Realities, Representations, Reflections*. Berkeley: International and Area Studies, 1999.

Zhuk, A. P. *Razvitie obshchestvenno-meditsinskoi mysli v Rossii v 60–70 gg. XIX veka*. Moscow: Gos. izd. meditsinskoi literatury, 1963.

Index

Academy, Medical Surgical, 125
Agapov, Andrei, 82, 83, 183, 184
Aleksandrovsk Textile Mill, 29
Alexander II, 110, 117
Altai, 21, 30, 34, 35, 38, 42, 48, 188n87
America, 63
Andreev, E. N, 3, 61, 71, 115, 131, 132
Andreev, Leonid, 127
apprenticeship, 8, 12, 13, 14, 25–30, 32, 33, 35, 40, 41, 56, 60, 94–95, 122, 186n12; abuse by employers of, 36, 41, 175; as a form of preparation for adulthood, 38, 45, 48; laws and regulations on, 28–32, 33, 56, 81, 176; perceptions of, 93, 104, 124, 174, 176; recruitment, 30; as reflected in literature, 126
Arkhiv, 124, 125
arteli, 63, 64, 73, 97, 99, 193n12
artisan workshops, 35, 58, 64, 67, 150, 184
Arutiunov, G. A., 143
associations, youth. *See* Budushchnost'; League of Youth, the; Northern Union of School Youth, the; South Russian Union of Youth, the; Yugenbund, the
associations of workers, voluntary. *See arteli*; *zemliazhestva*
Austria, 43
Azerbaijan, 140

backwardness, 145–46
Baklanova, E. N., 16

Balabanov, M., 61, 80, 144
Baranov, Count (Tver' governor), 104
Baranov, A. (a mill owner), 69
bast-matting factories, 65, 66, 71, 75, 76
Bernshtam, T. A., 24
Biblioteka dlia chteniia, 81
British cotton industry, 68
British Parliamentary Papers, 68
Budushchnost', 171
Bund, the, 171, 200n134
Bunge, N. Kh., 129, 132, 197n15
Building Code, 135
business arbitration courts, 97
Buzinov, Alexei, 152, 155

Chekhov, A. P., 1, 79, 80, 126, 185n2
child labor, 2, 4, 16, 21–23, 25, 27, 39, 49–51, 56–70, 76. *See also* apprenticeship; children; labor in industries; labor laws and regulations; working children
child rearing, 13–25, 125
childhood: popular and societal perceptions of, 15, 17–18, 20, 27, 94, 124, 127, 160, 176; as a scholarly concept, 3, 5–6, 15, 25, 113; state perceptions of, 27–28, 94, 124, 160; as a subject of contemporary interest, 124–26, 177
children: games and entertainment of, 20, 187n31; gendering of, 19; homelessness among, 159, 179; initiation in productive labor of, 7, 13, 20, 22, 25;

children: *(cont.)*
  involvement in labor protest and revolutionary activities, 167, 169, 171, 172, 178; in literature, 126, 136; in painting, 1; of serfs, 28, 31, 32, 35–37, 162; socialization of, 159; transition to adulthood, 23, 25. *See also* working children; labor; minimum employment age; employment of children; literature for children; mortality rates
"Children's Labor and Rest" (society), 165
cholera epidemic, 107
Christian morality, 114
civil society, 6, 109, 119, 177, 194n51, 195n86, 200n3
Cossacks, 171
Council of Machine-making Industrialists, 121
Crimean War, 94

Dement'ev, E. M., 66, 67, 72, 74, 79, 158, 185n11
demonstrations, 167–68, 169, 171
Denmark, 145
Dostoevsky, F. M., 1, 127, 179
dumas, 110, 119, 141
Dvinsk, 171

Ekaterinislav mill, 32
employment of children, 2–5, 9, 10, 13, 14, 27, 28, 31–33, 35, 36, 38–67, 69, 71, 77, 81, 84, 93–95, 97–102, 104–6, 108, 111, 113, 116, 117–32, 135–37, 140, 144–55, 159, 160, 162–64, 175, 176, 185n12, 188n68, 193n8, 193n12; causes of, 58, 62; in textiles, 49–51, 58, 62, 67–70, 76
England. *See* Great Britain
Erisman, F. F., 71, 131, 136
Europe, 15, 16, 24, 43, 44, 74, 95, 97, 137, 144, 145, 147
European Russia, 16, 47, 63, 73, 130, 148, 149

factories: manorial, 8, 12–13, 26–30, 32–36, 38–39, 42, 46, 174, 188n87, 189n1; state, 27, 30, 32, 38, 43
Factory and Apprenticeship Code, the, 94

factory inspectorate, 53, 108, 128, 130–32, 139, 141, 151, 153. *See also* factory inspectors
factory inspectors, 7, 43, 47, 55, 58, 67–69, 72–75, 80, 97–98, 107, 117, 124, 130–31, 133–36, 138, 140–41, 143–45, 147–53, 155, 157–58, 160, 163–64, 196n7, 196n10
factory labor laws. *See* labor laws and regulations
factory law specialists, 141, 95. *See also* Shtakel'berg, A. F.
factory physicians, 72, 124
factory schools, 113–16, 119, 121, 123, 138, 160–65
Fedoseev, Vasilii, 31
Filipov Candy Factory (Moscow), 155
Finance Ministry, 98, 108, 109, 111, 116, 123, 129, 130, 131, 134, 137; Department of Commerce of, 52; technical drawing schools of, 162
Finance Ministry Commission, 48, 59, 71, 94–97, 99, 106, 107, 109, 117. *See also* Ignat'ev's commission
Fink, D. F., 83
Finland, the Duchy of, 63
Fire Protection Code, 135
First Congress of Industrialists, 111
France, 2, 16, 43, 68, 97, 99, 144, 146, 185

gender scholarship, 10
generational conflict, 172
Gesse, S. G., 30, 33
Gessen, V. Iu., 4, 34, 57–59, 102
Georgia, 140
Gmelin J. G., 12
Goncharov textile mill, 33
Gorky, A. M., (Maxim) 1, 127, 185
Great Britain, 2, 3, 68, 144, 185
Guzhon factory (plant), 171
Guk factory, 85

Iakovlev Linen Mill, 32
Ianson, Iu. E., 120
Ianzhul I. I., 131
Iartsev Textile Mill, 52
Ignat'ev, P. N., 110
Ignat'ev's Comission, 110–11, 117, 122

Iletsk Salt Mines, 31
industrial districts: Kazan', 130, 149, 163; Kharkov, 130, 150, 163; Kiev, 130, 133, 157; Moscow, 130, 133, 136, 139, 163; St. Petersburg, 130, 133; Vilna, 130, 163; Vladimir, 58, 73, 130, 133, 163, 164; Voronezh, 130, 133 Warsaw, 130
industrialization, 2–5, 8, 10, 12–14, 24, 26, 36, 45–49, 56–57, 60, 63–65, 67, 70, 74, 79, 144, 146–47, 159, 170, 172, 174–75, 178, 189n3
industrial revolution, 2, 3, 24, 25
infants, 16–18
infanticide, 17
Interior Ministry, the, 40, 95, 98, 101, 106, 108, 116, 118, 122, 129, 139
Iartsev Textile Mill, 52
Iskra, 155
Ivanovo-Voznesensk, 119; Committee for Trade and Industry, 123

juveniles, 14, 16, 27, 31, 34, 36, 38, 48, 50, 52, 56, 65, 68, 71, 72, 83, 84, 96, 105, 111, 113, 121, 122, 125, 129, 130, 140, 142, 143, 154–56, 158, 159, 169

Kabuzan, V. M., 63
Kaigorodov, D. N., 113, 115, 123, 195n66
Kaluga province, 37, 187n39
Kankrin, E. F., 40
Kapnist, I. V., 41. *See also* Moscow: Province Civil Governor
Kazan', 34
Kelly, Catriona, 6, 186n16
Kokovtsov, N. M., 141
Kokovtsov's commission, 142
Korolenko, V. G., 127
Kozodavlev. O. P., 39
Krasnosel'skaia textile mill, 27
Kreenholm cotton mill, 4

labor in industries: in contemporary debates, 81, 94, 104, 105; impact on children's health, 120
labor laws and regulations: debates about, 6, 103, 112, 114, 124, 126–27, 177; early, 8, 12, 28, 39, 41, 42, 44–45, 94, 185n11; effectiveness of, 5, 146, 148–51; of 1861, 96, 98, 99, 107; of 1862, 98, 107; of 1882, 56, 129, 132, 135, 137, 139; of 1885, 138; of 1886, 138; of 1897, 148; in Europe, 43–44, 97, 122, 144–46; implementation of, 130, 137, 139, 141, 149, 150; of 1913, 143; of 1917, 159; pace and timing of, 43–44, 144–46, 159; popularization and publicity of, 124, 126, 144; reaction to, 100, 112–14, 139, 151; transgression of, 31, 136, 152–53
Land and Freedom (Zemlia I Volia) Society, 167
Laverychev V. A., 102, 193n19
League of Youth, the, 171
leather-tanning mills, 30, 32, 77
Lepeshkin (a merchant), 37. *See also* Voskresensk Cotton Mill
Lenin, V. I., 55, 64, 107, 147
Liadov, V. I., 125
literacy rates, 165. *See also* working children: literacy rates among
literature for children, 126–27
local economy, 14, 22, 23, 64

Marxist scholarship, 10, 186n17
match mills, 75, 76, 78, 81
McDouglass, M. L., 24
mechanization, 26, 46, 85
Medical Code, 135
Medical Surgical Academy, 125
Melancon, Michael, 4
men (adult males), 21, 24, 25, 53, 156, 165
metallurgical works, 12, 30, 34–5, 37–39, 53. *See also* Putilov plant; Guzhon factory (plant); Nizhne-Tagil'sk works
Mikhailov, N. F., 47, 48
Mikhailovskii, Ia. T., 55, 147, 163
Milogolova, I. I., 62
minimum employment age, 43, 102, 108, 116, 117, 121–23, 125, 131
mines, 1–2, 12, 28, 30–31, 33–35, 37–39, 42–43, 53, 98, 128, 130, 140, 148, 150, 156, 166, 188n87
Ministry for People's Education, 138
Mordvinov, N. S., Count, 27, 93
Morozov Cotton Mill (in Tver' province), 52, 74, 169

mortality rates: among infants and children in a comparative perspective, 16, 24; among rural infants and children, 17, 79; among rural population, 63, 79; in public discussions, 125; in urban centers, 79

Moscow, 23, 29, 39, 50, 51, 53, 58, 59, 66, 71, 75, 78, 82, 84, 98, 107, 110, 130, 131; Association for the Support of Industry, 139; city government commission on factory labor, 50; industrialists, 40, 100, 116, 133; industries, 50, 51, 59; military governor, 27, 41, 78; province, 23, 36, 45, 47–48, 66, 72, 107, 143, 161; Province Civil Governor, 41; Section of the Manufacturing Council, 40, 99, 100, 101, 108; State University, 131; Stock Exchange Committee, 118; textile mills, 30, 59

movement: labor, the, 10, 94, 110, 112, 113, 147, 167, 171, 185; political, 170; radical, 10, 144; revolutionary, 173; social-welfare reform, 124; socialist, 145; workers', 3, 110, 115, 116, 123, 143, 145, 166, 167, 185

Naidenov, N. A., 118
Nakhimov, P. S., 93
Nardinelli, Clark, 3, 103, 146, 185
Narymsk province, 21
Nekrasov, 1, 185
Nerchinsk mills, 31
Nevskii Cotton-Spinning Mill, 110, 168
Nicholas I, 40
Nikitin Cotton Mill, 64
Nikolaev, 110
Nizhne-Tagil'sk works, 12
Nizhnii Novgorod, 75, 187n39
Northern Union of School Youth, the, 171

Odessa, 110, 169, 171
Orel province, 19
Orenburg, 149

Pallas, P. S., 12–13
Pazhitnov, K. A., 35, 188n87

peasant migrants, 63
peasants, and peasantry, 5, 13, 15; labor duties of, 16; perceptions of life cycle among, 15, 17. *See also* peasant migrants
Penal Code, the, 135
Pereiaslavl'-Zalesskii Cotton Mill, 34
Perm' State Copper Works, 34
Peskov, P. A., 72–73, 76, 77, 131, 150–53, 164
Plekhanov, G. V., 166–67
Pod'iachii Metallurgical Works, 37
Poland, the Kingdom of, 63
Prokhorov, Konstantin, 161
Prokhorov, Timofei, 161
Proletarii, 155
Prussia, 43
Purlevskii, Savva, 17, 22, 23, 160
Putilov plant, 154

Rashin, A. G., 47, 190n8
Reforms of 1861–64, 26, 37, 38, 46, 60, 62, 64
Revolution, the: of February 1917, 158, 178; of 1905, 169, 171, 172; of October 1917, 4, 159, 178, 179
revolutionary movement, 173. *See also* workers' movement
Riazan', 29, 187n39
Riga, 110
Riga Stock Committee, 123
Rozhdestvensk Textile Mill (Tver' province), 52
Romanov, A., 58, 59
Russian Technical Society, 61, 125; Committee for Technical Education of, 49, 50, 67, 80, 120, 121, 131, 191n77
Russkaia mysl', 164
Russkie Vedomosti, 80

Samarin, D. F., 127
serfs, 23, 28, 31, 32, 35–37, 162, 186n19, 188n87
Serpukhov, 72
Shadrinsk district, 21
Shangina, I. I., 15
Shcherbatov, A. G., 41
Shtakel'berg, A. F., 48, 95

Shtakel'berg's commissions. *See* Finance Ministry Commission; St. Petersburg: commission
Siberia, 14, 12–13, 21–23, 31, 130, 166
Smolensk province, 20, 52
Snegirev ,V. S., 125
Society for the Support of Russian Industry and Commerce, 121, 123
Society for Workers' Welfare, 116
Sokolovskaia Cotton Mill, 68–69, 84, 164, 191n70
South Russian Union of Youth, 171
St. Petersburg, 39, 58, 59, 98, 110, 130, 133, 139, 149, 150, 153, 154, 157; commission, 95–96, 98, 107, 193n8, 193n18; Foundling Home of, 33; industrialists, 98, 100, 102, 193n19
State Council, Imperial, 27, 129, 132, 181
Statute on Rural Handicraft Workshops. *See* labor laws and regulations: of 1897
strikes, 4, 36, 37, 110, 120, 136, 143, 166–71; as addressed in labor laws, 128, 137, 142, 145; as a workers' protest instrument, 6
Sunday Schools, 161, 162
Suvorin, A. S., 127
Sviatopolk-Mirskii, P. D., 170
Sweden, 43

Tentelev Chemical Plant, 154
Third Department, 116
Thompson, E. P., 3, 185
Tolstoi, D. A., Count, 197n30
Tolstoy, L. N., 127
Tornton Mill, 169
tobacco factories, 49, 72, 111, 167
Torgovo-Pomyshlennaia Gazeta, 141
Torkovichi Glass Mill, 154
Tugan-Baranovsky, M. I., 41, 185n11
Tula entrepreneurs, 100, 101
Tula province, 183, 187n39
Tver' province, 52, 74, 100, 187n39

Ukraine, 21, 171
United States, the, 68, 180, 185n4
unemployment compensation, 141
Urals, the, 12, 31, 42

Valuev, P. A., 122–23
Vestnik Evropy, 86, 125, 130, 136, 144
Vladimir province, 23, 53, 64, 67, 76, 84, 104–5, 119–20, 150; Aleksandrovskii district (*uezd*) of, 69
Vologda province, 17–19, 186n19
Voronezh, 36
Voronin, M., 171
Voskoboinikov, A. A., 81, 85
Voskresensk Cotton Mill, 36, 41
Vreden, F. P., 113
Vyshnegradskii, I. A., 139

War Communism (1918–21), 159
Wigel Textile Factory, 36
Westernizers, 146
women (adult females), 23, 24, 53, 69, 138, 140, 147, 156, 158, 159, 165, 168; in labor law debates, 113, 114
women's employment, 44, 53, 68, 125, 137, 138, 140, 145
workday, 2, 5, 37, 40, 42–44, 70, 71, 76, 96, 99, 100, 102, 105–6, 108, 111–14, 117–20, 122–23, 128, 132, 137, 140, 155–57, 164, 176, 193n8; in comparative perspective, 44, 99, 142, 144–45
workers' morals: decline of, 170; in labor laws debates, 114. *See also* Christian morality
workers' movement. *See* movement: workers'
workers' protest, 6, 36, 108, 110, 136, 143, 166–69, 171, 172, 178. *See also* demonstrations; strikes; children: involvement in labor protest and revolutionary activities
working children: decline of health, 5, 9, 78, 86; diet of 73; education of, 160–65; living conditions of, 72, 73, 79, 186n5; literacy rates among, 163–65; recruitment of, 30; views on factory labor among, 77; wages of, 72, 76, 77, 157; work-related accidents among, 40, 77, 78, 82–85, 94, 95, 97, 103, 107, 108, 143, 156. *See also* employment of children
working conditions, 76, 78, 80, 84, 103, 147, 155, 172; as addressed in labor

working conditions, *(cont.)*
  regulations, 42, 130, 176; as addressed in scholarly debates, 3, 74
Workmen Compensation Act, the (Great Britain), 143
World War I, 145, 153, 154, 158, 173, 178

Yaroslavl' province, 17, 23, 187n39
Yugenbund, the, 170

Zakrevskii, A. A., 27
Zelnik, Reginald E., 4, 94
Zemliachestva, 97, 193n12
zemstvos, 110, 119, 141, 165; physicians, 79; schools, 162
Zhuk, A. P. (S. P. Lovtsov), 124, 125
Zmeinogorsk, 30
Zurich, 43